Test Bank

Bradley E. Garner
Boise State University

Carrie J. Garner

FUNCTIONS MODELING CHANGE

A PREPARATION FOR CALCULUS

Second Edition

Eric Connally
Harvard University Extension

Deborah Hughes-Hallett
University of Arizona

Andrew M. Gleason
Harvard University

et al.

WILEY

John Wiley & Sons, Inc.

CONTENTS

COVER PHOTO: ©Nick Wood

This material is based upon work supported by the National Science Foundation under Grant No. DUE-9352905. Opinions expressed are those of the authors and not necessarily those of the Foundation.

To order books or for customer service call 1-800-CALL-WILEY (225-5945).

ISBN 0-471-46827-4

Printed in the United States of America

10 9 8 7 6 5 4 3 2 1

Printed and bound by Bradford & Bigelow, Inc.

EXAM QUESTIONS for FUNCTIONS MODELING CHANGE

Chapter 1 Exam Questions

Exercises

1. Write the relationship of population, P, of a city as a function of time, t, in years using function notation.
   ANSWER:
   $P = f(t)$

2. Do the graphs below give y as a function of x (YES or NO)? If NO, say why.

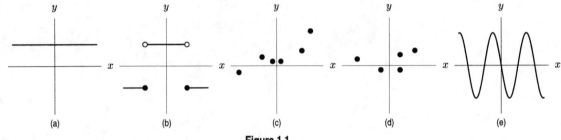

Figure 1.1

   ANSWER:
   (a) Yes.
   (b) Yes.
   (c) Yes.
   (d) No, two different values of y for one x.
   (e) Yes.

3. Use Table 1.1 to fill in the missing values. (There may be more than one answer.)
   (a) $p(40) =$?   (b) $p(?) = 40$ (c) $p(20) =$?   (d) $p(?) = 20$

Table 1.1

s	0	10	20	30	40
$p(s)$	0	20	30	40	20

   ANSWER:
   (a) 20   (b) 30   (c) 30   (d) 10; 40

4. Let $y = p(x)$ be defined by the graph in Figure 1.2.

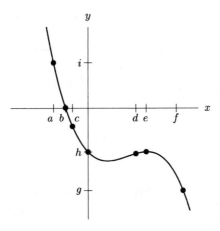

Figure 1.2

(a) Evaluate $p(a)$ and $p(0)$.
(b) Solve the following equation for x: $p(x) = 0$.
(c) Is $p(\frac{1}{2}d)$ closer to g, h, or i?
(d) Is $p(-b)$ positive or negative or equal to zero?
(e) Solve $p(x) \geq 0$ for x.

ANSWER:

(a) $p(a) = i$; $p(0) = h$
(b) If $p(x) = 0$, $x = b$.
(c) The value $p(\frac{1}{2}d)$ is closest to h.
(d) $p(-b)$ is negative.
(e) If $p(x) \geq 0$, then $x \leq b$.

5. Label the axes for a sketch to illustrate the statement "graph the number of units sold, q, as a function of the price, p."
ANSWER:

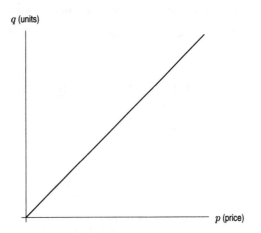

Figure 1.3

4

6. On what intervals is the function graphed in the following figure increasing? Decreasing?

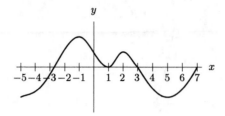

Figure 1.4

ANSWER:

Increasing from -5 to -1, 1 to 2, and 5 to 7; decreasing from -1 to 1 and 2 to 5.

7. Table 1.2 shows the height of a tree (to the nearest foot) as a function of time (in years), What was the average rate of change in the height of the tree between

   (a) $t = 1$ and $t = 3$?
   (b) $t = 4$ and $t = 6$?
   (c) $t = 10$ and $t = 12$?
   (d) Interpret these results in terms of the height of the tree.

Table 1.2

t	0	1	2	3	4	5	6	7	8	9	10	11	12
height	5	5	6	7	9	11	13	14	14	15	15	15	15

ANSWER:

(a) $\dfrac{7-5}{3-1} = \dfrac{2}{2} = 1$

(b) $\dfrac{13-9}{6-4} = \dfrac{4}{2} = 2$

(c) $\dfrac{15-15}{12-10} = \dfrac{0}{2} = 0$

(d) The tree grew slowly at first, then more quickly, then essentially stopped growing.

8. Calculate the average rates of change for the function $f(x) = -x^2 + 2x$ between $x = -1$ and $x = 0$ and between $x = 1$ and $x = 2$.

ANSWER:

$$\frac{f(0) - f(-1)}{0 - (-1)} = \frac{0 - (-3)}{1} = 3$$

$$\frac{f(2) - f(1)}{2 - 1} = \frac{0 - 1}{1} = -1$$

9. Figure 1.5 shows the graph of the function $y = f(x)$. Estimate $\Delta f(x)$, Δx, and the rate of change, $\Delta f(x)/\Delta x$, for the following values:

(a) $x = 5$ and $x = 10$

(b) $x = 20$ and $x = 25$

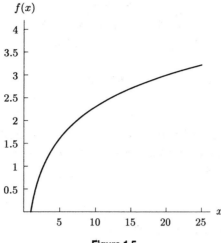

Figure 1.5

ANSWER:

(a) $\Delta f(x) \approx 2.3 - 1.6 = 0.7$; $\Delta x = 10 - 5 = 5$; $\Delta f(x)/\Delta x \approx 0.7/5 = 0.14$.

(b) $\Delta f(x) \approx 3.2 - 3.0 = 0.2$; $\Delta x = 25 - 20 = 5$; $\Delta f(x)/\Delta x \approx 0.2/5 = 0.04$.

10. Could the following table represent a linear function?

Table 1.3

t	0	1	2	3	4
$p(t)$	1000	1400	1600	1700	1750

ANSWER:

No. The average rate of change is not constant. For example, $\dfrac{1400 - 1000}{1 - 0} = 400$, while $\dfrac{1600 - 1400}{2 - 1} = 200$.

11. Could the following table represent a linear function?

Table 1.4

x	2	4	6	8	10
$f(x)$	160	130	100	70	40

ANSWER:

Yes. The average rate of change is -15 in each case.

12. Could the following table represent a linear function?

Table 1.5

t	0	1	3	5	10
$g(t)$	0.6	0.95	1.65	2.35	4.1

ANSWER:

Yes. The average rate of change is 0.35 in each case.

13. A salesman earns $W(n) = 300 + 75n$, where n is the number of products sold, and $W(n)$ is the weekly wage, in dollars. Identify the vertical intercept and the slope, and explain their meaning in practical terms.

   ANSWER:

   Vertical intercept is 300, which is the salesman's base salary. Slope is 75, which is the commission for each product sold.

14. The distance traveled by a car is $D(t) = 65t$, where t is the number of hours driven, and $D(t)$ is the distance in miles. Identify the vertical intercept and the slope, and explain their meaning in practical terms.

   ANSWER:

   Vertical intercept is 0, which is the distance traveled when the car began the trip. Slope is 65, which is the speed of the car in miles per hour.

15. If possible, rewrite $4x + 3y + 8 = 2x$ in slope-intercept form, $y = b + mx$.

   ANSWER:

   $3y = -8 - 2x$

16. Find a formula for the linear function with slope 1/2 that passes through the point (-1,5).

   ANSWER:

   Using the formula $y = b + mx$,

   $$5 = b + \frac{1}{2}(-1)$$
   $$b = \frac{11}{2}.$$

   Thus, the line is $y = \frac{11}{2} + \frac{1}{2}x$.

17. Find a formula for the linear function with $y-$intercept -1 that passes through the point (2,-3).

   ANSWER:

   Using the formula $y = b + mx$,

   $$-3 = (-1) + m(2)$$
   $$-2 = 2m$$
   $$m = -1.$$

   Thus, the line is $y = -1 - x$.

18. Table 1.6 gives data from a linear function. Find a formula for the function.

   Table 1.6

Time, t	0	1	2
Population, $P = f(t)$	50,000	40,000	30,000

   ANSWER:

   Since $f(0) = 50,000$, the $y-$intercept is 50,000. Use any two points to find the slope:

   $$\frac{40,000 - 50,000}{1 - 0} = -10,000.$$

   Thus, $P = 50,000 - 10,000t$.

19. Find equations for the three lines shown in Figure 1.6. Your equations will involve the constants given on each graph. Be sure to simplify your answers.

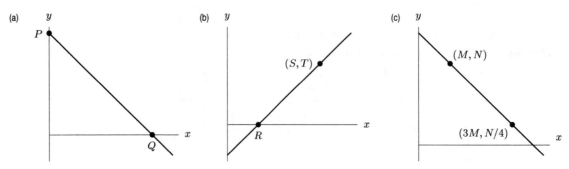

Figure 1.6: The graphs of three different linear functions

ANSWER:

(a) The y-intercept is $y = P$. The slope is

$$m = \frac{\Delta y}{\Delta x} = \frac{P - 0}{0 - Q} = -\frac{P}{Q}.$$

Thus,

$$y = P - \frac{P}{Q}x.$$

(b) The slope is

$$m = \frac{\Delta y}{\Delta x} = \frac{T - 0}{S - R} = \frac{T}{S - R}.$$

The line passes through the point $(R, 0)$ and so, since $y = b + mx$, we have

$$0 = b + \left(\frac{T}{S - R}\right)R$$

and so

$$b = -\frac{RT}{S - R}.$$

Thus,

$$y = \frac{T}{S - R}x - \frac{RT}{S - R},$$

which can be written as

$$y = \frac{Tx - RT}{S - R}.$$

(c) The slope is

$$m = \frac{N - (N/4)}{M - 3M} = \frac{(3/4)N}{-2M} = -\frac{3N}{8M}.$$

Thus,

$$y = b - \frac{3N}{8M}x.$$

Since the line passes through the point (M, N) we have

$$N = b - \frac{3N}{8M}(M)$$
$$N = b - \frac{3}{8}N$$

and so

$$b = N + \frac{3}{8}N$$
$$b = \frac{11}{8}N.$$

Thus,

$$y = \frac{11}{8}N - \frac{3N}{8M}x.$$

20. Six linear functions are defined for all values of x by the equations below.

(i) $\frac{1}{3}f(x) - 1 = x$ (ii) $-\frac{1}{3}g(x) = x + 1$ (iii) $2h(x) + x = 3$
(iv) $j(x) = -3x$ (v) $k(x) = 3$ (vi) $2p(x) + x = 5$

Without giving reasons, state which of these functions

(a) are decreasing.
(b) have graphs which are parallel.
(c) have graphs which pass through the point $(0, 3)$.
(d) are neither increasing nor decreasing.

ANSWER:

First, we rewrite the functions as follows:

(i) $f(x) = 3x + 3$

(ii) $g(x) = -3x - 3$

(iii) $h(x) = -\frac{1}{2}x + \frac{3}{2}$

(iv) $j(x) = -3x$

(v) $k(x) = 3$

(vi) $p(x) = -\frac{1}{2}x + \frac{5}{2}$

This should make it a little easier to see that

(a) $g(x), h(x), j(x), p(x)$ are decreasing.

(b) $h(x)$ and $p(x)$ are parallel; $g(x)$ and $j(x)$ are parallel.

(c) $f(x)$ and $k(x)$ pass through the point $(0, 3)$.

(d) $k(x)$ is neither increasing nor decreasing, it is a horizontal line.

21. (a) Suppose that $f(x)$ is linear and that $f(-2) = 4$, $f(5) = -3$. Find a formula for f.

(b) Suppose that f is linear with slope $-\frac{2}{3}$ passing through the point $(1, 8)$. Find a formula for f.

(c) Suppose that f is linear, with an x-intercept of -7 and a y-intercept of -2. Find its formula.

(d) Suppose that f is linear with x-intercept=2, and parallel to $y = -4x + 2$. Find its formula.

ANSWER:

(a) $\frac{-3-4}{5-(-2)} = \frac{-7}{7} = -1$. So our line is of the form $y = -x + b$. Substituting $(-2, 4)$ into this equation yields $4 = 2 + b; b = 2$. Therefore $y = -x + 2$.

(b) Using the point slope form of a line we have $y - 8 = \frac{-2}{3}(x - 1)$.

(c) Since $(-7, 0)$ and $(0, -2)$ are on the line $m = \frac{0-(-2)}{-7-0} = \frac{-2}{7}$. Therefore $y = \frac{-2}{7}x - 2$.

(d) The slope of the line we seek is -4 and $(2, 0)$ is on this line. Therefore $y = -4(x - 2)$.

22. Find an equation for the line that satisfies each of the following conditions:

(a) Passes through $(-3, -5)$ and $(2, 7)$,

(b) Passes through $(-5, 4)$ with slope $\frac{2}{3}$,

(c) Has x-intercept of -4 and y-intercept of 3,

(d) Is parallel to $y = \frac{4}{9}x + 8$ and passes through $(9, -16)$,

(e) Is perpendicular to $y = -\frac{3}{7}x + 2$ and passes through $(-3, 5)$,

(f) Has slope $-\frac{1}{2}$ and forms the hypotenuse of a triangle lying in Quadrant I whose area is 3 and whose legs lie on the x and y axes.

ANSWER:

(a) $m = \frac{7-(-5)}{2-(-3)} = \frac{12}{5}$ so $y = 7 + \frac{12}{5}(x - 2)$ so $y = \frac{12}{5}x + \frac{11}{5}$

(b) $y = 4 + \frac{2}{3}(x - (-5))$ so $y = \frac{2}{3}x + \frac{22}{3}$

(c) $m = \frac{3}{4}$ so $y = \frac{3}{4}x + 3$

(d) $m = \frac{4}{9}$ so $y = -16 + \frac{4}{9}(x - 9)$ so $y = \frac{4}{9}x - 20$

(e) $m = \frac{-1}{(\frac{-3}{7})}$ so $y = 5 + \frac{7}{3}(x - (-3))$ so $y = \frac{7}{3}x + 12$

(f) Suppose the x- and y-intercepts of the line are a and b respectively. Then we get $\frac{b}{a} = \frac{1}{2}$, hence $a = 2b$. So the area of the triangle is $\frac{ab}{2} = 3$, hence $b^2 = 3$. So $b = \sqrt{3}$ and the equation of the line is $y = \sqrt{3} - \frac{1}{2}x$, so $y = -\frac{1}{2}x + \sqrt{3}$.

23. Figure 1.7 gives lines $A, B, C, D,$ and E. Without a calculator, match each line to $f, g, h, j,$ or k. $f(x) = 3 + 2x$
$g(x) = 3 - 2x$
$h(x) = 3 + \dfrac{1}{2}x$
$j(x) = 3 - \dfrac{1}{2}x$
$k(x) = 3$

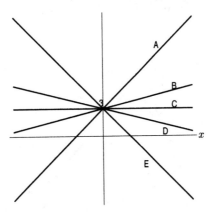

Figure 1.7

ANSWER:
$A, f; \ B, h; \ C, k; \ D, j; \ E, g$

24. Match the r values with the scatter plots in the figures:

$$r = 0.9, \ r = 0.5, \ r = 0, \ r = -0.5, \ r = -0.9$$

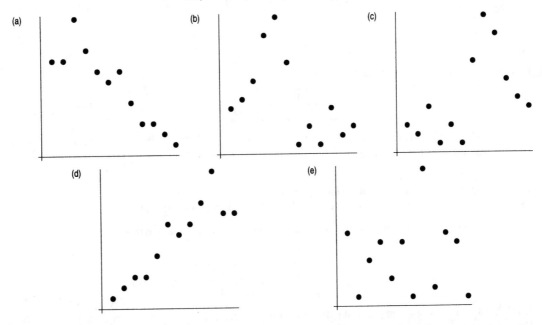

ANSWER:

(a) -0.9
(b) -0.5
(c) 0.5
(d) 0.9
(e) 0

25. Table 1.7 shows a store's total sales, $S(p)$, of an item when it is priced at price p, in dollars.

    (a) Make a scatter plot of this data.

    (b) Draw a regression line by eye.

    (c) Roughly estimate the correlation coefficient by eye.

Table 1.7

p	10	11	12	13	14	15
$S(p)$	25	23	20	18	17	15

ANSWER:

(a)

(b)

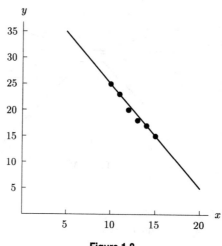

Figure 1.8

    (c) Roughly -0.8

Problems

26. Consider the information in the table:

s	3	0	1	−1	6	2	7
t	−2	5	3	1	2	5	11

    (a) Could t be a function of s? Explain.

    (b) Could s be a function of t? Explain.

    ANSWER:

(a) Yes. Any given s value determines a unique t value.

(b) No. If $t = 5$, s could be 0 or 2. So not every value of t determines a unique value of s.

27. Suppose that Table 1.8 shows the cost of a taxi ride, in dollars, as a function of of miles traveled.

Table 1.8

m	0	1	2	3	4	5
$C(m)$	0	2.50	4.00	5.50	7.00	8.50

    (a) What does $C(3.5)$ mean in practical terms? Estimate $C(3.5)$.

    (b) If $C(m) = 3.5$ what does m mean in practical terms? Estimate m.

    ANSWER:

(a) $C(3.5)$ represents the cost of a 3.5 mile taxi ride. Since a three mile taxi ride costs $5.50 and a four mile taxi ride costs $7.00 a 3.5 mile taxi ride will cost $6.25.

(b) m represents the number of miles one can travel if one pays the taxi driver $3.50. Since $2.50 will get you one mile and $4.00 will get you two miles, $3.50 will get you about 1.66 miles.

28. If S is a surface area of a sphere of radius r, and V is its volume, then $S = 4\pi r^2$, and $V = \frac{4}{3}\pi r^3$. Find V as a function of S.

ANSWER:

From the expression for S, we can solve for r, obtaining $r = \sqrt{\frac{S}{4\pi}}$. Now we substitute this expression for r into the equation for V, which yields $V = \frac{4}{3}\pi \left(\sqrt{\frac{S}{4\pi}} \right)^3$. Simplifying this expression gives $V = \frac{1}{6}\sqrt{\frac{S^3}{\pi}}$.

29. Match each story about a person's weight to one of the graphs, where w represents the person's weight and t is time in months.

(a) Starts out at 150 pounds and loses 5 pounds a month.
(b) Starts out at 150 pounds and loses 10 pounds a month.
(c) Starts out at 140 pounds and remains at that weight.
(d) Starts out at 140 pounds and gains 5 pounds a month.

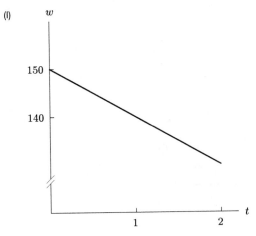

Figure 1.9

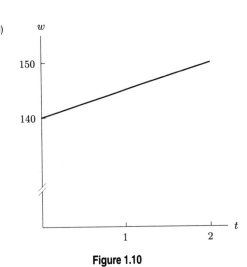

Figure 1.10

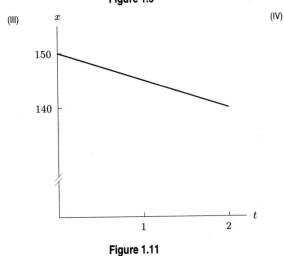

Figure 1.11

Figure 1.12

ANSWER:

(a) (III) (b) (I) (c) (IV) (d) (II)

30. Sketch a graph that could represent the average daily temperature of New York City over a one-year period as a function of time, t, with $t = 0$ being January 1.

ANSWER:

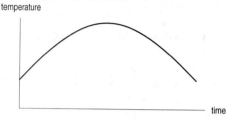

Figure 1.13

31. Many economists believe that a country's rate of inflation will rise if its unemployment rate falls, Figure 1.14 shows the relationship between the inflation rate, I, and the unemployment rate, u, in four different countries.

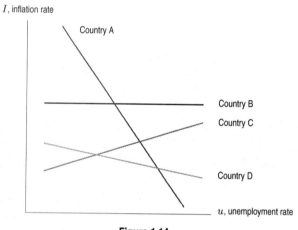

Figure 1.14

(a) In which of the four countries does the relationship between I and u seem to be the opposite of what is nomally expected?

(b) In which of the four countries does a change in the unemployment rate have no apparent affect on the inflation rate?

(c) In which of the four countries is the inflation rate most sensitive to change in the unemployment rate?

ANSWER:

(a) For country C, the rate of inflation rises as the unemployment rate *rises*. This is the opposite of what was expected.

(b) For country B, the inflation rate stays constant as the unemployment rate changes. Thus, the unemployment rate appears not to affect the inflation rate.

(c) For country A, small decreases in the unemplopyment rate correspond to the most drastic increases in the inflation rate. Thus, country A's inflation rate is the most sensitive to change in the unemployment rate.

32. The Kentucky Derby is a horse race which takes place once a year. The horses which enter are generally male; female horses usually compete in the Kentucky Oaks which takes place the day before. In May 1995 a female horse named Serena's Song entered in the Derby and was favored to win. For most of the race she led; however near the end she was overtaken and so did not end up winning. Sketch a possible graph showing the distance of Serena's Song and the winning horse from the starting point as a function of to time.

ANSWER:

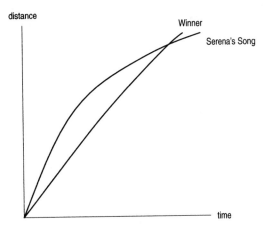

Figure 1.15

33. Suppose $f(x) = x^2$.

(a) Find the average rate of change of the function f between $x = 1$ and $x = 4$.

(b) Find the value of c making the average rate of change between $x = 1$ and $x = c$ twice your answer to part (a).

ANSWER:

(a) For $f(x) = x^2$ between $x = 1$ and $x = 4$

Average rate of change $= \dfrac{f(4) - f(1)}{4 - 1} = \dfrac{4^2 - 1^2}{4 - 1} = \dfrac{15}{3} = 5.$

(b) Between $x = 1$ and $x = c$

$$\text{Average rate of change} = \frac{f(c) - f(1)}{c - 1}$$
$$= \frac{c^2 - 1^2}{c - 1} = \frac{(c + 1)(c - 1)}{c - 1} = c + 1.$$

If this average rate of change is twice the answer to part (a), we have

$$c + 1 = 2(5)$$
$$c = 9.$$

34. Match each of the descriptions below with the single most appropriate graph. Each graph may be used more than once or not at all.

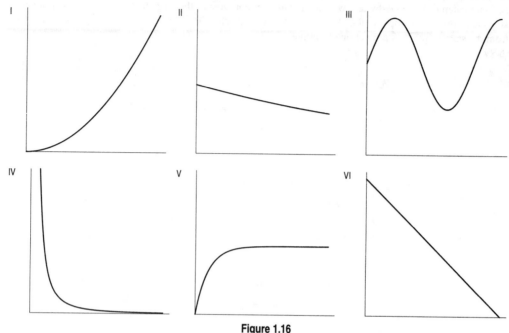

Figure 1.16

(a) The average daily temperature in Kansas City as a function of the day of the year.

(b) The gravitational force between two particles as a function of the distance between them. Newton's law of gravitation says that the gravitational force is proportional to the inverse square of the distance.

(c) The area of a circle as a function of radius.

(d) The quantity of radioactivity in a sample decaying as a function of time.

(e) The number of bacteria in a sample with an unlimited food supply and unlimited space.

(f) The resale price of a car which depreciates steadily until it is worthless.

(g) The inventory stocked in a warehouse where new stock arrives at a faster rate than sales take place, but the difference between the restocking rate and sales is decreasing.

(h) The inventory stocked in a warehouse where sales outpaces restocking at a fixed rate.

ANSWER:

(a) Since the average daily temperature of Kansas City will cycle with the slowly changing seasons, we can expect the graph to matched III.

(b) Since gravitational force $g = \dfrac{k}{\text{distance}^2}$, we can expect the graph to drop sharply from a very high y value when x is close to 0, to $y \approx 0$ when at a large x value. This is best matched by graph IV.

(c) Since area of a circle $= \pi r^2$, this is best matched by graph I.

(d) We can expect the graph to have a finite y-intercept and to decline. This is best matched by graph II.

(e) Since the number of bacteria in a sample with unlimited food supply grows at a constant percent rate. Thus, we can expect the graph to show exponential growth, which is best matched by graph I.

(f) A car with steady depreciation until it is worthless can be described by a linear function with negative slope. This is best described by graph VI.

(g) Since the restocking is always faster than the sales, we can expect the graph to have a positive slope. But because the difference in rates is decreasing, we know the graph will have a correspondingly decreasing slope. This is best matched by graph V.

(h) Since sales outpaces stocking at a constant rate, we can expect the graph to be linear with negative slope. This is best matched by graph VI.

35. Sketch a graph to depict the following scenario: After recovering from an injury, an athlete begins training again for the marathon. As she gets stronger, she is able to run faster. She runs a full marathon once a week. Sketch a graph of the number of minutes it takes her to run a marathon as a function of the number of weeks since her recovery from the injury. Label axes.

ANSWER:

Answers vary. First, we know that the graph will only make sense for positive x. Second, since the athlete gets stronger and runs faster, the time it takes her to run the marathon gets shorter the longer she has been training. Thus we can say that the graph must depict a decreasing function. We also know that it will never take the athlete a negative amount of minutes to run the marathon (or even 0 minutes for that matter) We will assume that the graph will level off to her pre-injury time. Thus the graph of the amount of minutes it takes the athlete to run the marathon as a function of the number of weeks she has been training will look like the following figure:

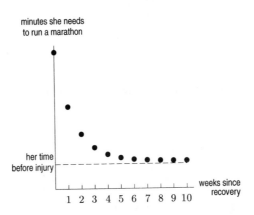

Note: The graph shows that the athlete does not attain her pre-injury marathon time within ten weeks. An alternative answer would be to show the graph dipping below its horizontal asymptote, showing that she is in fact able to beat her previous time. This choice is left to the interpretation of the student sketching the graph.

36. A car company has found that there is a linear relationship between the amount of money it spends on advertising and the number of cars it sells. When it spends 50 thousand dollars on advertising, it sold 500 cars. Moreover, for each additional 5 thousand dollars spent, they sell 20 more cars.

(a) Let x be the amount money they spend on advertising, in thousands of dollars. Find a formula for y, the number of cars sold.

(b) What is the slope of your equation? What is the meaning of the slope in terms of the problem?

ANSWER:

(a) Since y is a linear function, we can write

$$y = mx + c$$

where m and c are constants to be determined when $x = 50,000$, $y = 500$

$$500 = 50,000m + c$$

when $x = 55,000$, $y = 520$

$$520 = 55,000m + c$$

so

$$500 - 50,000m = 520 - 55,000m$$
$$5000m = 20$$
$$m = \frac{20}{5000} = 0.004.$$
$$c = 500 - (50000)(0.004)$$
$$= 300.$$
$$y = 0.004m + 300.$$

(b) The slope of the equation is m, the coefficient of x, which is equal to 0.004. It is the number of cars sold per dollar spent on advertising.

37. One of the functions represented by data in the table below is linear. Determine which of the functions is linear and find a formula to represent the function.

t	g(t)	h(t)
1	22.40	2.2
2	25.09	2.5
3	28.10	2.8
4	31.47	3.1
5	35.25	3.4
6	39.48	3.7

ANSWER:

In general, if a function $f(t)$ is linear, then the difference quotient of *any* two distinct values t_1 and t_2 is constant, equal with the slope of f:

$$\frac{f(t_2) - f(t_1)}{t_2 - t_1} = m = \text{constant}.$$

In our case we notice that only $h(t)$ has this property:

$$\frac{2.5 - 2.2}{2 - 1} = \frac{2.8 - 2.5}{3 - 2} = \cdots \frac{3.7 - 3.4}{6 - 5} = 0.3$$

Thus, $h(t)$ is linear with slope 0.3, so $h(t) = 0.3 \cdot t + b$. To find b, we know from the table that for $t = 1$, $h(t) = 2.2$. Thus, $2.2 = 0.3(1) + b$, so $b = 1.9$. Then formula for h is:

$$h(t) = 0.3t + 1.9.$$

38. ComElectric, the Cambridge power company, charges its customers $7.50 a month and 6.885¢ per kwh, plus a 3.129¢ per kwh surcharge. (A kwh, or kilowatt-hour, is a unit of electricity supplied.)

(a) Write a formula giving your monthly cost, C, in dollars, if you use x kwh of electricity a month.

(b) Sketch C against x for $x > 0$. Label intercepts.

ANSWER:

(a) We know that

$$\text{Cost} = \text{Monthly charge} + \text{Power charge} \cdot \text{Power used} + \text{Surcharge} \cdot \text{Power used}.$$

Since the monthly charge = $7.50, the power charge = 6.885¢, the surcharge = 3.129¢, and the power used = x, we have,

$$C = \$7.50 + 6.885¢x + 3.129¢x$$
$$= \$7.50 + 10.01¢x$$

Since the answer is to be given in dollars, we must convert the cents to dollars using 100¢ = $1.

$$C = \$7.50 + 10.01¢(\$1/100¢)x = 7.50 + 0.10x$$

(b)

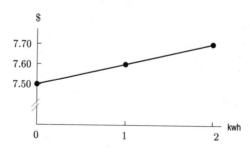

Figure 1.17

39. The cost, C, of producing q items is given by the formula

$$C = C_0 + mq$$

where C_0, m are positive constants. Circle the correct answer in each of the questions below.

(a) If the quantity produced decreases, does the cost of production:

- Decrease
- Stay the same
- Increase
- Can't tell without knowing values of C_0 and m

(b) If the quantity produced doubles, does the cost of production:

- More than double
- Exactly double
- Less than double
- Can't tell without knowing values of C_0 and m

ANSWER:

(a) Decreases. We know that

$$C = C_0 + mq.$$

Thus, if q decreases, cost of production decreases because C_0 and m are both constants. When production goes down, cost goes down. See Figure 1.18.

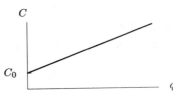

Figure 1.18

(b) Less than double. If production is doubled, the cost will rise but will be less than double. This is because if the cost of production were to double,

$$2 \cdot C = 2 \cdot (C_0 + mq)$$

then C_0 would also have to double

$$2C = 2C_0 + 2mq$$

Since only q is doubling and not C_0, the actual cost of production will be less than double.

40. Find possible formulas for the following linear population functions.

(a) This population is 80,000 in year $t = 0$ and grows by 250 people per year.

(b) This population is 60,000 in year $t = 8$ and 78,000 by year $t = 20$.

ANSWER:

(a) $P(t) = 80,000 + 250t.$

(b) Since the population function is linear, the rate of change of the population is given by:

$$m = \frac{\Delta P}{\Delta t} = \frac{78,000 - 60,000}{20 - 8} = \frac{18,000}{12} = 1500 \text{ people/year.}$$

Therefore, the linear function is

$$P(t) = 60,000 + 1500(t - 8)$$
$$P(t) = 48,000 + 1000t.$$

41. An airplane has room for 300 coach-fare seats. It can replace any 3 coach seats with 2 first-class seats. Suppose the airplane is configured with x coach seats and y first-class seats. (Assume no space is wasted.)

(a) If $y = 0$, what is the value of x?

(b) If $x = 0$, what is the value of y?

(c) It can be shown that y is a linear function of x. Sketch a graph of y versus x, labeling both axis-intercepts.

(d) Find a formula for y in terms of x.

(e) Explain the significance of the slope of the formula you found in part (d) in terms of the airplane's seat configuration.

(f) Explain the significance of the x and y-intercepts in terms of the airplane's seat configuration.

ANSWER:

(a) If there are no first class seats, there must be 300 coach seats.

(b) If there are no coach seats there must be 200 first class seats

(c)

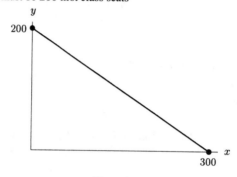

Figure 1.19

(d) $m = \frac{-200}{300}$ and $b = 200$, so $y = 200 - \frac{2}{3}x$.

(e) The slope tells us the replacement value for each group of 3 coach seats removed in terms of first class seats that can be added.

(f) The y-intercept is the configuration of a plane that has no coach seats while the x-intercept is the configuration of a plane that has no first class seats.

42. At a price of $1.30 per gallon, the average weekly demand by consumers for gasoline is 42 gallons. If the price rises to $1.35 per gallon, the weekly demand drops to 39 gallons.

(a) Find the formula for Q, the weekly quantity of gasoline demanded, in terms of p, the price per gallon, assuming that demand is linear.

(b) What is the economic significance of the slope of your formula?

(c) What are the Q- and p-intercepts for your formula? Interpreted literally, what is the economic interpretation of the Q-intercept and the p-intercept? Are these interpretations realistic?

ANSWER:

(a)

$$Q = b + mp$$

$$m = \frac{\Delta Q}{\Delta p} = \frac{39 - 42}{1.35 - 1.30} = \frac{-3}{0.05} = -60$$

So, using $Q = 42$ for $p = 1.30$, we have

$$Q = Q_0 + m(p - p_0)$$
$$Q = 42 - 60(p - 1.30) \quad \text{(point-slope)}$$

or

$$Q = 120 - 60p \quad \text{(slope-intercept)}$$

(b) The slope is -60 gallons/dollar. This means that an increase of $1/gallon in gasoline prices will result in a 60 gallon decrease in weekly demand.

(c) The Q-intercept of $Q = 120$ tells us that at a price of $0/gallon, 120 gallons would be demanded every week. The p-intercept is the value of p if $Q = 0$:

$$120 - 60p = 0$$
$$60p = 120$$
$$p = 2.$$

Thus, at a price of $2/gallon, no gasoline will be demanded. These interpretations are not very realistic. It is unlikely that the price of gasoline would ever be $0/gallon, or that at a price of $2/gallon there would be no demand whatsoever.

43. (a) Write the equation of a line with x-intercept of x_0 and y-intercept of y_0.

(b) Is your answer to part (a) unique? (Assuming that x_0 and y_0 are fixed). Explain.

ANSWER:

(a) The equation for a line is $y = b + mx$ where $m =$ slope and $b = y$-intercept. Since we know that $y_0 = y$-intercept and that

$$\text{Slope} = \frac{\text{rise}}{\text{run}} = \frac{y_0 - 0}{0 - x_0} = -\frac{y_0}{x_0}$$

we have

$$y = \left(-\frac{y_0}{x_0}\right)x + y_0.$$

(b) Unless $x_0 = y_0 = 0$, the answer to part (a) is unique because two points $(x_0, 0)$ and $(0, y_0)$ define a line.

44. An athlete wanting to strengthen his cardiovascular system will bench-press a weight, w, as many times, N, as possible. The data in Table 1.9 show the relation between N and w.

Table 1.9

Weight, w (pounds)	140	150	160	170
Max number bench presses, N	29	26	23	20

(a) Graph N against w.

(b) Assuming linearity, express N as a function of w.

(c) What are the physical interpretations of the slope and vertical intercept?

(d) What is the maximum weight this athlete will be able to lift if he bench-presses once?

ANSWER:

(a)

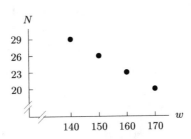

Figure 1.20

(b) The slope is

$$\frac{29 - 26}{140 - 150} = -0.3 \text{ number of lifts/pound,}$$

so the equation is of the form

$$N = b - 0.3w.$$

Substituting $N = 26$ and $w = 150$ gives

$$26 = b - 0.3(150)$$
$$b = 71$$

and therefore

$$N = 71 - 0.3w.$$

(c) The slope, -0.3 bench-presses/pound, represents the decrease in number of bench-presses that the athlete can do as the weight increases by 1 pound. However, bench-presses are integer units so this might better be expressed as a decrease of three bench-presses for each additional 10 pound increase. The vertical intercept, 71, is the number of bench-presses that the athlete could do if the weight is zero. (The linear relationship between N and w probably does not hold down to $w = 0$, however.)

(d) The maximum weight the athlete can bench-press is given by the value of w when $N = 1$:

$$1 = 71 - 0.3w$$
$$w = \frac{-70}{-0.3} = 233.3 \text{ pounds.}$$

45. Find the equation of the line l in Figure 1.21 in terms of the constants A and B.

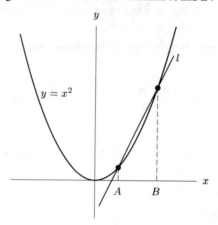

Figure 1.21

ANSWER:

Using the two points (A, A^2) and (B, B^2), we get

$$\text{slope} = \frac{A^2 - B^2}{A - B} = \frac{(A - B)(A + B)}{A - B} = A + B.$$

Thus,

$$y - A^2 = (A + B)(x - A)$$
$$y - A^2 = (A + B)x - A^2 - AB$$
$$y = (A + B)x - AB.$$

46. If $A = 65 + 1.7t$ and $B = 26 + 4.1t$ represent the populations (in millions) of two countries, A and B, then in what year will

(a) The population of country A equals that of country B?

(b) Country A has 30 million more inhabitants than B?

(c) Country A has twice as many inhabitants as B?

ANSWER:

(a) If $A = B$, then $65 + 1.7t = 26 + 4.1t$. Thus, $39 = 2.4t$ and $t = 16.25$.

(b) If $A = B + 30$, then $65 + 1.7t = 25 + 4.1t + 30$. Thus, $9 = 2.4t$ and $t = 3.75$.

(c) If $A = 2B$, then $65 + 1.7t = 2(26 + 4.1t) = 52 + 8.2t$. Thus $13 = 6.5t$ and $t = 2$.

47. **(a)** Find the equation of line l, shown in Figure 1.22.

(b) Find the coordinates of point P.

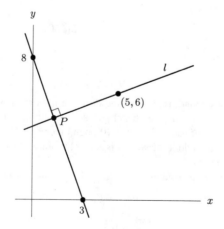

Figure 1.22

ANSWER:

(a) First, we find the equation of the line perpendicular to l. We have

$$m = \frac{8-0}{0-3} = -\frac{8}{3}.$$

Since $(0,8)$ is on this line, this gives $b = 8$, and so

$$y = 8 - \frac{8}{3}x.$$

Thus, the slope of line l is given by

$$m = \frac{-1}{-8/3} = \frac{3}{8}.$$

Since a point on line l is $(5,6)$, we have

$$y = 6 + \frac{3}{8}(x-5) \quad \text{or} \quad y = \frac{33}{8} + \frac{3}{8}x.$$

(b) Point P is the intersection of $y = 8 - \frac{8}{3}x$ and $y = 6 + \frac{3}{8}(x-5)$. This gives

$$8 - \frac{8}{3}x = 6 + \frac{3}{8}(x-5)$$
$$8 - 6 = \frac{8}{3}x + \frac{3}{8}x - \frac{15}{8}$$
$$\frac{64}{24}x + \frac{9}{24}x = 2 + \frac{15}{8}$$
$$\frac{73}{24}x = \frac{31}{8}$$
$$x = \left(\frac{31}{8}\right)\left(\frac{24}{73}\right)$$
$$x = \frac{93}{73}.$$

Thus,

$$y = 8 - \frac{8}{3}\left(\frac{93}{73}\right)$$
$$= 8\left(1 - \frac{31}{73}\right)$$
$$= 8\left(\frac{42}{73}\right)$$
$$= \frac{336}{73}$$

We have

$$P = (93/73, 336/73) \approx (1.274, 4.603).$$

48. Using the window $-10 \le x \le 10, -10 \le y \le 10$, graph the equations $y = x$, $y = 0.5x$, $y = 0.1x$, and $y = 0.01x$ on the same set of axes.

(a) Explain what happens to the graphs of the lines as the slope becomes small.

(b) Explain why the fourth line does not appear in the window.

ANSWER:

(a) The lines get closer to the $x-$axis until they are nearly horizontal.

(b) The fourth line is so close to the $x-$axis that you cannot distinguish them in this window.

49. You need to purchase a new computer printer. The cost of printer A is \$75, and its ink cartridges are \$28 each. The cost of printer B is \$120, and its ink cartridges are \$18 each. You estimate that you will need new ink cartridges about once a month.

(a) Write equations for the total cost $C_A = f(t)$, for printer A and its ink and $C_B = g(t)$, for printer B and its ink, where t is time in months.

(b) Graph both of these on the same set of axes.

(c) After about how many months would you have invested the same amount of money in each?

ANSWER:

(a) $C_A = f(t) = 75 + 28t;\ C_B = g(t) = 120 + 18t.$

(b)

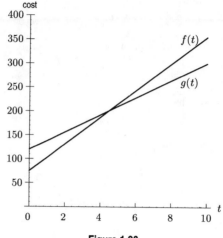

Figure 1.23

(c) About 4.5 months (the point where the lines cross).

50. The wild rabbits of Australia have recently been seriously threatened by a virus that was accidentally released into their population. (The rabbits themselves were released into Australia decades ago, and have become serious pests. The decimation by the virus is not being viewed by many as a bad thing.) Suppose that Table 1.10 gives the number r of rabbits (in millions) remaining t months after the release of the virus.

Table 1.10

t (months)	2	4	6	8	10	12	14	16	18	20
r (millions)	1940	1842	1649	1328	1140	898	765	502	296	104

(a) Find the equation of the least-squares regression line for this data. (Be sure to use t and r in the equation, rather than x and y.)

(b) Do you believe that the linear model found in (a) fits the data well? Justify your answer.

(c) Use the linear model found in (a) to estimate the rabbit population 15 months after the release of the virus. Round your answer to the nearest million rabbits, and be sure to include units.

(d) What is the physical interpretation of the t-intercept of the graph of the line found in (a)? Your answer should be stated in real-world terms involving rabbits and months, rather than in abstract mathematical terms.

(e) Would you expect this model to be useful to estimate the rabbit population before time $t = 0$? What about after time $t = 22$? Give reasons; the more specific, the better.

ANSWER:

(a) By using a calculator program, we can find the regression line

$$r = -106.097t + 2213.47.$$

(b) The correlation coefficient is $-.997744$, which is nearly -1. Thus, the regression line is an excellent model for the given data.

(c)

$$r = -106.097(15) + 2213.47 \approx 622 \text{ million rabbits}$$

(d) The physical interpretation of the t-intercept is that it is the time in months at which the rabbit population falls to zero. (The rabbits will become extinct during month 20.)

(e) This model would not be useful before $t = 0$ because at that point, the rabbits were not affected by the virus. Thus, the premise of the function would be incorrect at that time interval. Similarly, after time $t = 22$, this model is also not very effective, as it would suggest that the population is negative, which goes contrary to the physical realities.

51. A study was done that collected data on 100 individals' grade point averages and the number of hours per week they spent watching televsion. A linear regression was done on this data and the corresponding correlation coefficient was −0.95. What does that tell you about the relationship between grade point averages and television watching?

ANSWER:

The correlation coefficient of −0.95 tells us that the data more or less lie on a line with negative slope, that is, the more television is watched, the more the grade point average drops.

52. Figures 1.24 and 1.25 give USDA data on milk cows. Figure 1.24 shows N, the number of milk cows (in millions) as a function of t, the number of years since 1986. Figure 1.25 shows r, the annual amount of milk produced (in 1000s of pounds) per cow, also as a function of t.

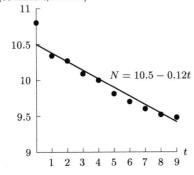

Figure 1.24: US Milk Cows (millions)

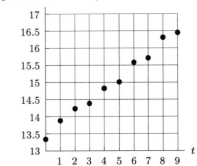

Figure 1.25: Rate of milk production (1000s lbs/cow)

(a) In the graph at left, a regression line has been drawn to fit the data, and its formula is shown. Is it reasonably correct?

(b) Fit a regression line to the points in the other graph and find its formula. You can either use a calculator to find the line of best fit, or you can fit the regression line "by eye."

ANSWER:

(a) Yes

(b) Using a calculator, we find that

$$r = 0.34t + 13.4.$$

[Note: your answer may differ somewhat.]

Chapter 2 Exam Questions

Exercises

1. Evaluate $f(x) = x^2 + x + 1$ for $x = -2$.
   ANSWER:
   $$f(-2) = (-2)^2 + (-2) + 1 = 4 - 2 + 1 = 3$$

2. Solve $f(x) = \sqrt{x+1} = 3$ for x.
   ANSWER:

$$\sqrt{x+1} = 3$$
$$x + 1 = 3^2 = 9$$
$$x = 8$$

3. If $f(x) = \dfrac{1}{x+1} + 3$,

   (a) Find $f(0)$
   (b) Solve $f(x) = 0$.

   ANSWER:

   (a)
   $$f(0) = \frac{1}{0+1} + 3 = 1 + 3 = 4$$

   (b)
   $$\frac{1}{x+1} + 3 = 0$$
   $$\frac{1}{x+1} = -3$$
   $$1 = -3(x+1) = -3x - 3$$
   $$3x = -1 - 3 = -4$$
   $$x = -\frac{4}{3}$$

4. $A = f(s) = s^2$ is the area of a square with side length s. Evaluate $f(4)$, solve $f(s) = 4$ (for $s \geq 0$), and interpret your answers in terms of the square .
   ANSWER:
   $f(4) = 4^2 = 16$, which is the area of a square with side 4. $f(s) = 4$ means $s^2 = 4$, so $s = 2$. This means a square with area 4 has side length 2.

5. (a) Find the domain of $r(x) = \dfrac{1}{(x+2)^2} + \sqrt{1-x}$.

   (b) Find the range of $s(x) = \dfrac{6-x}{3+x}$.

   ANSWER:

   (a) The function r is defined at a particular value of x provided the functions $1/(x+2)^2$ and $\sqrt{1-x}$ are both defined for that value of x. The function $1/(x+2)^2$ is defined when $x \neq -2$. The function $\sqrt{1-x}$ is defined if $1 - x$ is not negative, that is, if

   $$1 - x \geq 0$$
   $$1 \geq x.$$

   Thus the domain of r is all values of x such that $x \leq 1$ and $x \neq -2$.

(b) If y is a number in the range of s, then

$$y = \frac{6 - x}{3 + x}$$

for some value of x. Solving for x, we have

$$y(3 + x) = 6 - x$$
$$3y + xy = 6 - x$$
$$x + xy = 6 - 3y$$
$$x(1 + y) = 6 - 3y \quad \text{Factoring}$$
$$x = \frac{6 - 3y}{1 + y} \quad \text{provided } y \neq -1.$$

Thus we see that the range of $s(x)$ is all y such that $y \neq -1$.

6. Find domain and range of the function

$$f(x) = \frac{1}{\sqrt{1 - x^2}}.$$

ANSWER:

$-1 < x < 1$ is the domain for $f(x)$. The range of $f(x)$ is all $y \geq 1$.

7. The graph of $g(x) = \sqrt{9 - x^2}$ is shown in Figure 2.1. The range of $g(x)$ is (circle one):

(a) $y \geq 0$ **(b)** $-3 \leq y \leq 3$ **(c)** $y \leq 3$ **(d)** $0 \leq y \leq 3$

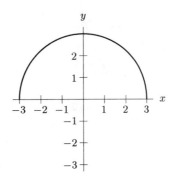

Figure 2.1

ANSWER:

(d)

8. Estimate the domain and range of $f(x)$. Assume the entire graph is shown.

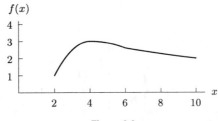

Figure 2.2

ANSWER:

The domain is $2 \leq x \leq 10$. The range is $1 \leq f(x) \leq 3$.

9. Evaluate $h(3)$ if $h(x) = \begin{cases} 3x + 4 & \text{for } x \leq 2 \\ -x^2 + 1 & \text{for } x \geq 2 \end{cases}$

    (a) 13   **(b)** 10   **(c)** -8   **(d)** 3

    ANSWER:

    (c) -8

10. Graph the piecewise defined function $f(x)$. Use an open circle to represent a point which is not included and a solid dot to indicate a point which is on the graph.

$$f(x) = \begin{cases} 1 & \text{if } -2 \leq x < 0 \\ x^2 & \text{if } 0 \leq x < 2 \\ x & \text{if } 2 \leq x \leq 4 \end{cases}$$

    ANSWER:

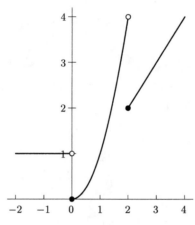

Figure 2.3

11. Write a formula for the function $y = f(x)$.

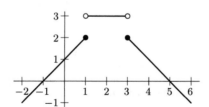

Figure 2.4

    ANSWER:

$$f(x) = \begin{cases} x + 1 & \text{if } x \leq 1 \\ 3 & \text{if } 1 < x < 3 \\ -x + 5 & \text{if } x \geq 3 \end{cases}$$

12. Let $f(x) = x^2 + 3x + 2$. Use the graph of $y = f(x)$ to estimate $f^{-1}(50)$. Explain your method.

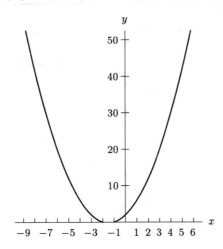

Figure 2.5

ANSWER:

From the sketch we can read off an estimate of the value of x for which $f(x) = 50$. $f^{-1}(50) = 5.6$, or -8.6.

13. Let f be given by the graph in Figure 2.6. Evaluate the following quantities:

(a) $f(-3)$
(b) $f^{-1}(2)$
(c) $f(2)$
(d) $f^{-1}(-3)$

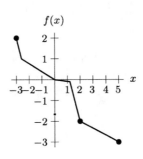

Figure 2.6

ANSWER:

(a) 2
(b) −3
(c) −2
(d) 5

14. Table 2.1 gives values of an invertible function, f..

(a) $f(1) = ?$ (b) $f(?) = 1$ (c) $f^{-1}(1) = ?$ (d) $f^{-1}(?) = 1$

Table 2.1

x	0	1	2	3	4
$f(x)$	-1	0	1	3	5

ANSWER:

(a) 0   (b) 2   (c) 2   (d) 0

15. The surface area of a sphere of radius r is given by $A = f(r) = 4\pi r^2$. Find a formula for the inverse function, $f^{-1}(A)$, giving radius as a function of surface area.

ANSWER:

Solve $A = 4\pi r^2$ for r :

$$A = 4\pi r^2$$
$$\frac{A}{4\pi} = r^2$$
$$r = \sqrt{\frac{A}{4\pi}}.$$

16. Calculate successive rates of change for the function $g(t)$, in Table 2.3 to decide whether you expect the graph of $g(t)$ to be concave up or concave down.

Table 2.2

t	0	1	2	3
$g(t)$	-1	0	2	8

ANSWER:

The successive rates of change are $\dfrac{0 - (-1)}{1 - 0} = 1$, $\dfrac{2 - 0}{2 - 1} = 2$, and $\dfrac{8 - 2}{3 - 2} = 6$. Since these numbers are increasing, the graph of the function would probably be concave up.

17. Calculate successive rates of change for the function $g(t)$, in Table 2.3 to decide whether you expect the graph of $g(t)$ to be concave up or concave down.

Table 2.3

x	0.1	0.2	0.3	0.4
$f(x)$	-2.3	-1.6	-1.2	-0.9

ANSWER:

The successive rates of change are -7, -4, and -3 Since these numbers are increasing, the graph of the function would probably be concave up.

18. Does the graph of the following function appear to be concave up, concave down, or neither?

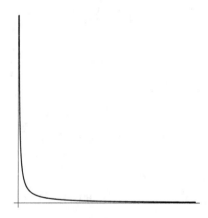

Figure 2.7

ANSWER:

Concave up

19. Does the graph of $y = x^2 - x$ appear to be concave up, concave down, or neither?

ANSWER:

Concave up

20. Find the zeros of $Q(r) = 2r - 3 + r^2$ by factoring.

ANSWER:

$$2r - 3 + r^2 = 0$$
$$r^2 + 2r - 3 = 0$$
$$(r + 3)(r - 1) = 0$$

Thus, the zeros are at $r = -3$ and $r = 1$.

21. Find the zeros of $y = 6x^2 + 14x + 4$ algebraically.

ANSWER:

Set $y = 0$ and solve by factoring:
$$6x^2 + 14x + 4 = 0$$
$$(3x + 1)(2x + 4) = 0$$

Thus, the zeros are at $x = -1/3$ and $x = -2$.

22. Find two quadratic functions with zeros $x = -1$ and $x = 2$..

ANSWER:

One function would be $(x + 1)(x - 2) = x^2 - x - 2$. Another function would be a scalar multiple of this, such as $2x^2 - 2x - 4$ or $3x^2 - 3x - 6$.

23. Solve $2.5x^2 - 3x = 0.5$ using the quadratic formula.

ANSWER:

Rewrite the equation as $2.5x^2 - 3x - 0.5 = 0$, and use the quadratic formula with $a = 2.5$, $b = -3$, and $c = -0.5$. Thus

$$x = \frac{-(-3) \pm \sqrt{(-3)^2 - 4(2.5)(-0.5)}}{2(2.5)} = \frac{3 \pm \sqrt{14}}{5}.$$

Thus, the roots are $x = \dfrac{3 + \sqrt{14}}{5}$ and $x = \dfrac{3 - \sqrt{14}}{5}$.

Problems

24. Suppose that your uncle Gerald has just opened a store that sells television sets. Let $P(x)$ represent the profit that uncle Gerald makes by selling x television sets per week. Suppose that uncle Gerald currently sells k television sets per week. Using complete sentences, state in everyday real-world terms, involving profit and television sets, what each of the following means. Do not use the symbols for x, k, or P directly in your answers. Thus, for example, if you had to refer to $P(k)$, you would instead write "the profit uncle Gerald is making at his current level of sales".

(a) $P(k) + 1000$

(b) $P(k + 5)$

(c) $P(2k)$

ANSWER:

(a) This formula represents the profit uncle Gerald is making at his current level of sales plus one thousand dollars.

(b) This formula represents the profit uncle Gerald would make if he sold five more television sets per week over his current level of sales.

(c) This formula represents the profit uncle Gerald would make if he sold twice as many television sets per week as his current level of sales

25. Given $g(x) = x^2 - x$, what is $g(1 - x)$?

(a) $x^2 - x$ (b) $x - x^2$ (c) $x^2 - 3x$ (d) $-x^2 - x$

ANSWER:

(a) $x^2 - x$

26. For $f(x) = x^2 - 3x + 4$, evaluate and simplify:

(a) $f(10)$

(b) $f(x + 10)$

(c) $f\left(\frac{1}{x-2}\right)$

ANSWER:

(a)

$$f(10) = 10^2 - 3 \cdot 10 + 4$$
$$= 100 - 30 + 4$$
$$= 74$$

(b)

$$f(x + 10) = (x + 10)^2 - 3(x + 10) + 4$$
$$= x^2 + 20x + 100 - 3x - 30 + 4$$
$$= x^2 + 17x + 74$$

(c)

$$f\left(\frac{1}{x-2}\right) = \left(\frac{1}{x-2}\right)^2 - 3\left(\frac{1}{x-2}\right) + 4$$
$$= \frac{1}{x^2 - 4x + 4} - \frac{3}{x-2} + 4$$
$$= \frac{1}{x^2 - 4x + 4} - \frac{3}{x-2}\frac{x-2}{x-2} + 4\frac{x^2 - 4x + 4}{x^2 - 4x + 4}$$
$$= \frac{1}{x^2 - 4x + 4} - \frac{3x - 6}{x^2 - 4x + 4} + \frac{4x^2 - 16x + 16}{x^2 - 4x + 4}$$
$$= \frac{1 - 3x + 6 + 4x^2 - 16x + 16}{x^2 - 4x + 4}$$
$$= \frac{4x^2 - 19x + 23}{x^2 - 4x + 4}$$

27. Suppose $g(x) = x^2 - 2x$.

    (a) Evaluate $g(x + h)$.

    (b) Evaluate and simplify $\frac{g(x+h) - g(x)}{h}$.

      ANSWER:

    (a) To obtain an expression for $g(x + h)$, we substitute $(x + h)$ for x in the expression for $g(x)$:

$$g(x + h) = (x + h)^2 - 2(x + h)$$
$$= x^2 + 2xh + h^2 - 2x - 2h$$
$$= x^2 + (2h - 2)x + (h^2 - 2h)$$

    (b)

$$\frac{g(x + h) - g(x)}{h} = \frac{(x + h)^2 - 2(x + h) - (x^2 - 2x)}{h}$$
$$= \frac{x^2 + 2hx + h^2 - 2x - 2h - x^2 + 2x}{h}$$
$$= \frac{2hx + h^2 - 2h}{h}$$
$$= \frac{h(2x + h - 2)}{h}$$
$$= 2x + h - 2$$

28. Let $w(m)$ give the weight (in pounds) of an average-sized baby girl who is m months old.

    (a) An average six-month old girl weighs 15 pounds. Express this fact using the function $w(m)$.

    (b) What does the statement $w(12) = 21$ tell you about baby girls? Be specific.

    (c) What does the statement $w(p) = 14$ tell you about p? Be specific.

    (d) An eight-month old baby girl named Amelia weighs $w(13)$ pounds. Is Amelia of average weight, above average weight, or below average weight?

ANSWER:

(a) $w(6) = 15$

(b) The statement tells us that an average 12-month old girl weighs 21 pounds.

(c) The statement tells us that p is the age in months of an average girl weighing 14 pounds.

(d) Amelia, who is only 8 months old, weighs as much as an average 13-month old baby. Thus, she is of above average weight.

29. Let $I(x)$ be the federal income tax paid based on a total taxable income of x dollars. For 1996, the value of $I(x)$ is represented by the graph in Figure 2.8.

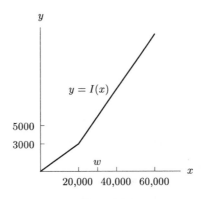

Figure 2.8

(a) Which of the following would represent an increase in your income tax if your income was w dollars? Circle all that apply.
$$I(w + 500), \qquad I(w) - 200, \qquad I(w - 100), \qquad I(w) + 400$$

(b) In in this context which is greater: $I(x + 500)$ or $I(x) + 500$?

ANSWER:

(a) Circle the first and last

(b) The value of $I(x) + 500$

30. Assume that height is a function of age and that $H = f(a)$ is the average height (in inches) for females in the US at age a (in years).

(a) Write brief statements to explain what the following expressions represent:
  (i) $f(1.5)$
  (ii) $f(z) + 5$
  (iii) $f(0)$

(b) Determine a reasonable domain and range for $H = f(a)$.

ANSWER:

(a) (i) $f(1.5)$ gives the average height of a female who is 1.5 years old
   (ii) $f(z) + 5$ is 5 inches taller than the average height of a female z years old.
   (iii) $f(0)$ is the average height of newborn female in the US.

(b) Domain: $0 \leq a \leq 110$. Range: $20 \leq H \leq 66$.
   Note: Your answers may vary! Explanations are important here. Possible reasons are:
   The domain must be $a \geq 0$ (notice that 0 makes sense) but also there must be some upper bound. We could put 110, as few US females live beyond 110 years. The range is the set of outputs. We estimated newborns at about $20''$ (average) and average adults at about $5'6''$.

31.

Table 2.4

s	-1	0	1	2	3	4	5	6
t	2	1	2	3	2	5	5	2

The data points from Table 2.4 are graphed in both Figures 2.9 and 2.10. Use the graphs to answer the following questions.

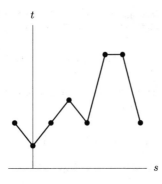

Figure 2.9

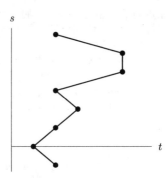

Figure 2.10

(a) Which graph represents the graph of a function?

(b) Fill in the following with either a t or an s: (You do not need to find a formula for f!)
For the graph that is a function, we could use the notation _____ = $f($_____$)$, where _____ is the dependent variable and _____ is the independent variable.

(c) Use the graph of the function to approximate $f(1.5)$.

(d) What is the domain and range of the function f?

(e) Give a brief explanation of why the other graph does not represent a function.

ANSWER:

(a) Figure 2.9 represents the graph of a function.

(b) $t = f(s)$, where t is the dependent variable and s is the independent variable.

(c) $f(1.5) \approx 2.5$.

(d) The domain of the function is $-1 \le s \le 6$, and the range of the function is $1 \le t \le 5$.

(e) Answers may vary. Figure 2.10 does not represent a function, because for some given independent variable t, there exists more than one dependent variable s.

32. An organization has 100 raffle tickets to sell for $1.00 each. From the proceeds of the sale, they will award the winning ticket holder $25.00. The profit, P, the organization makes is a function of the number of tickets, n, sold. Find the domain and range of this function and sketch its graph.

ANSWER:

The domain is $0 \le n \le 100$. The range is $-25 \le P \le 75$.

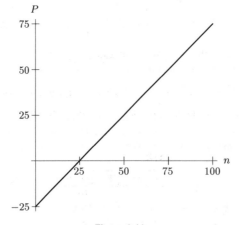

Figure 2.11

33. Find the domain and range of $h(x) = \dfrac{1}{\sqrt{x+a}}$, where a is a constant.

ANSWER:

$h(x)$ is defined where $x + a > 0$, so the domain is $x > -a$. The range is all positive real numbers, $h(x) > 0$.

34. Let $f(x) = \dfrac{1}{x + |x|}$. What is the domain of f? Explain.

ANSWER:

If $x \le 0$, then $|x| = -x$ and so $|x| + x = 0$, but if $x > 0$ then $|x| = x$ and so $|x| + x = 2x$. Thus $f(x) = \frac{1}{2x}$ for $x > 0$. But $f(x) = \frac{1}{0}$, i.e. undefined for $x \le 0$ and so the domain is $x > 0$. (In other words, you get a zero in the denominator for any $x \le 0$.)

35. ComGas, the Cambridge power company, charges its customers $6.56 a month, plus an additional 87.357¢ per therm up to 28 therms or 57.987¢ per therm above 28 therms. (A therm is a unit of gas supplied.)

(a) Express C, your monthly cost in dollars, as a function of x, the number of therms of gas used in a month.

(b) Sketch C against x for $x > 0$. Put scales on the axes and label intercepts.

ANSWER:

(a) If $0 \le x \le 28$

$$\text{Cost} = \text{Monthly cost} + 87.357¢ \cdot \text{Conversion to dollars} \cdot x$$
$$= \$6.56 + \frac{87.357¢}{100¢} \cdot x$$
$$= \$6.56 + \$0.87357x.$$

If $x > 28$

$$C = \text{Monthly cost} + 87.357¢ \cdot \text{Conversion to dollars} \cdot 28 + 57.987¢ \cdot \text{Conversion to dollars} \cdot (x - 28)$$
$$= \$6.56 + \frac{87.357¢}{100¢} \cdot 28 + \frac{57.987¢}{100¢}(x - 28)$$
$$= \$31.02 + \$0.57987(x - 28)$$

Thus we write $f(x) = \begin{cases} 6.56 + 0.87357x & \text{for } 0 \le x \le 28 \\ 31.02 + 0.57987 \cdot (x - 28) & \text{for } x > 28 \end{cases}$

(b)

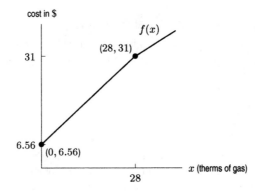

Figure 2.12

36. A T-shirt printing company charges a set-up fee of $10 for each order, plus the cost per shirt shown in Table 2.5.

(a) Express C, the total cost in dollars, as a piecewise function of n, the number of shirts ordered.

(b) Sketch a graph of C for $0 \le n \le 40$.

Table 2.5

# of shirts	cost per shirt
0-10	$10
11-20	$9
21-30	$8
over 30	$7

ANSWER:

(a)

$$C = \begin{cases} 10 + 10n & \text{if} \quad 0 \le n \le 10 \\ 10 + 9n & \text{if} \quad 10 < n \le 20 \\ 10 + 8n & \text{if} \quad 20 < n \le 30 \\ 10 + 7n & \text{if} \quad n > 30 \end{cases}$$

(b)

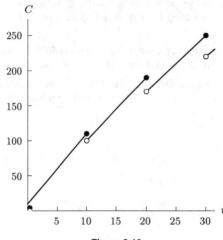

Figure 2.13

37. The function $D(p) = 1200 - 200p$ gives the weekly demand for video rentals at Joe's Videolog when Joe charges p dollars to rent a video. What does $D^{-1}(50)$ mean in terms of the problem?

    ANSWER:

    $D^{-1}(50)$ is the rental fee Joe must charge to get a weekly demand of 50 video rentals.

38. Give the meaning and units of the inverse function $f^{-1}(H)$ where $H = f(t)$ is the average height, in inches, of a child t years old.

    ANSWER:

    $f^{-1}(H)$ is the average age, in years, of a child H inches tall.

39. The circumference, in cm, of a circle whose radius is r cm is given by $C = 2\pi r$.

    (a) Write this formula using function notation, where f is the name of the function.
    (b) Evaluate and interpret $f(r + 2)$.
    (c) Evaluate and interpret $f(r) + 2$.
    (d) Evaluate and interpret $f^{-1}(8\pi)$.

    ANSWER:

    (a) $C = f(r) = 2\pi r$.
    (b) $f(r + 2) = 2\pi(r + 2) = 2\pi r + 4$, the circumference of a circle of radius $r+2$.
    (c) $f(r) + 2 = 2\pi r + 2$, 2 more than the circumference of a circle of radius r.
    (d) $f^{-1}(8\pi) = 4$, the radius of a circle with circumference 8π.

40. The number of cancer cells grows slowly at first but then grows with increasing rapidity. Draw a possible graph of the number of cancer cells against time.

    ANSWER:

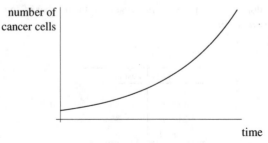

41. Each of the four stories below describes how some quantity is varying as a function of time. In each case, state the quantity the function describes and sketch a graph of it against time.

   (a) Even though the child's temperature is still rising, the penicillin seems to be taking effect.

   (b) The cost of a new car is increasing at an ever increasing rate.

   (c) The price of memory chips isn't decreasing as quickly it used to be.

   (d) The annual profits made by Upper Midwest Industries is decreasing at a higher rate each year.

     ANSWER:

   (a) Function: temperature of child.

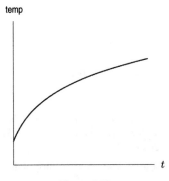

Figure 2.14

   (b) Function: cost of car.

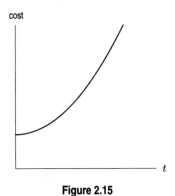

Figure 2.15

   (c) Function: price of memory chip.

Figure 2.16

(d) Function: profit of UMI.

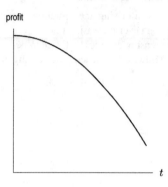

Figure 2.17

42. The probability of being in an accident increases as a driver's blood-alcohol content (BAC) rises. It has been observed that the probability rises faster and faster as the BAC increases. Let $P(x)$ be the probability that a driver will be involved in an accident as a function of x, the driver's BAC.

   (a) Sketch a graph of $P(x)$ and label the axes. You should clearly indicate whether or where your graph is increasing, decreasing, concave up, or concave down.

   (b) Let L be the BAC at or above which it is illegal to drive. Describe in words the meaning of the following expressions in the context of drinking and driving.

   (i) $P(0)$ (ii) $P(L)$ (iii) $P(2L)$

   (c) Let $b = P(0)$. Describe in words the meaning of the expression $P^{-1}(5b)$ in the context of drinking and driving.

   ANSWER:

   (a) The graph in Figure 2.18 is everywhere increasing and concave up.

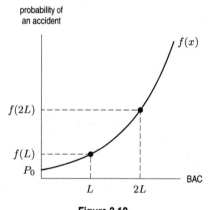

Figure 2.18

   (b) (i) $P(0)$ is the probability of an accident for someone who has not been drinking.

   (ii) $P(L)$ is the probability of an accident for someone who is at the legal BAC limit.

   (iii) $P(2L)$ is the probability of an accident for someone whose BAC is twice the legal limit.

   (c) If $b = P(0)$, then b is the probability of an accident for someone who hasn't been drinking. Since P gives a probability as a function of a BAC, we see that P^{-1} gives a BAC as a function of probability. Thus, $P^{-1}(5b)$ is a BAC — specifically, the BAC at which the probability of an accident is 5 times that of a person who hasn't been drinking.

43. On the axes below, sketch a graph of a continuous function, $y = f(x)$ with all of the following features:

- $f(0) = -2$
- $f(-2)$ and $f(3) = 0$
- f is decreasing for $x < 0$
- f is increasing for $x > 0$
- f is concave up for $x < 3$
- f is concave down for $x > 3$
- $f(x) \to \infty$ as $x \to -\infty$
- $f(x) \to 4$ as $x \to \infty$

ANSWER:

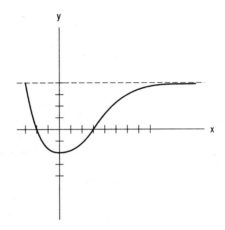

44. Each of the statements (a) - (e) describes a quantity changing with respect to time; the quantity is in parentheses after the statement. For each of the statements (a) - (e), choose one of the graphs (i)- (vi) and one of the statements (vii) - (xi).

(a) Energy is being used at an ever increasing rate. (Total energy used)

(b) The car slowed down steadily. (Velocity)

(c) The rumor spread — first slowly, then faster, and then slower until everyone had heard it. (Number of people who have heard the rumor)

(d) The pollution decreased, but more slowly as time went on. (Quantity of pollution)

(e) The quantity of the drug in the patient's blood stream increased at a decreasing rate until it reached saturation. (Quantity of drug in blood)

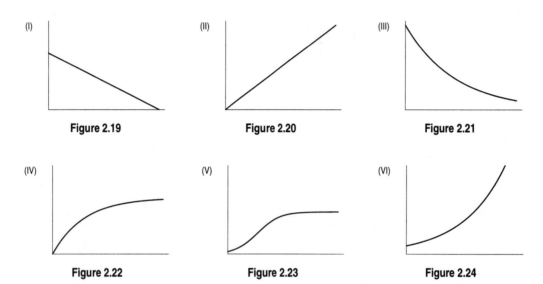

| (I) Figure 2.19 | (II) Figure 2.20 | (III) Figure 2.21 |
| (IV) Figure 2.22 | (V) Figure 2.23 | (VI) Figure 2.24 |

   (vii) Increasing and concave up

 (viii) Decreasing and concave up

   (ix) Increasing and concave down

    (x) Decreasing and concave down

   (xi) None of (vii) - (x)

ANSWER:

(a) Graph VI, statement vii.
(b) Graph I, statement xi.
(c) Graph V, statement xi.
(d) Graph III, statement viii.
(e) Graph IV, statement ix.

45. Determine the concavity of the graph of $f(x) = x^2 + x - 4$ between $x = -2$ and $x = 1$ by calculating average rates of change over intervals of length 1.

ANSWER:

The average rate of change over the first interval is given by

$$\frac{f(-1) - f(-2)}{-1 - (-2)} = \frac{-4 - (-2)}{1} = -2.$$

Similarly, the average rates of change over the next two intervals are 0 and 2. Since these numbers are increasing, the function is concave up.

46. Graph a quadratic function which has all of the following properties: concave up, y−intercept at 8, zeros at $x = -3$ and $x = -1$.

  ANSWER:

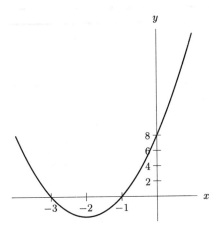

Figure 2.25

47. A model rocket is launched from the roof of a building. For height h, in meters, and time t, in seconds, after the rocket is launched, the height on the rocket above the ground is given by

$$h = f(t) = -4.9t^2 + 42t + 20.$$

See Figure 2.26

(a) Find and interpret the domain and range of the function and the intercepts of the graph.

(b) Identify the concavity.

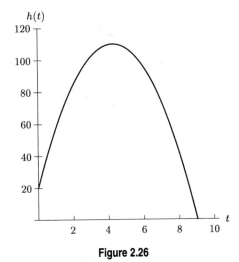

Figure 2.26

  ANSWER:

(a) To find the domain, we need to know at what time the rocket hits the ground. We find this be solving $h = f(t) = -4.9t^2 + 42t + 20 = 0$ using the quadratic formula. This gives $t \approx 9.024$. The domain is the time the rocket is in the are: $0 \le t \le 9.024$. To find the range, we look for the smallest and largest values of h. Looking at the graph, these are approximately 0 and 110. Thus, the range is $0 \le h \le 110$. The $h-$ intercept, 20, is the rockets initial height, or the height of the building. The $t-$ intercept, 9.024, is the time it takes for the rocket to hit the ground.

(b) The graph is concave down.

Chapter 3 Exam Questions

Exercises

1. In 2002, the cost of a piece of computer equipment was \$230 and going down at a rate of 9% per year. Assuming this percentage remains constant, express the cost C, in dollars, of this equipment as a function of time, t, the number of years since 2002.

ANSWER:

$C = 230(0.91)^t$

2. You start with \$1000. How much money will you have after each effect?

(a) 10% increase

(b) 25% decrease

(c) 20% increase followed by 20% decrease

ANSWER:

(a) $1000 + 1000(0.10) = \$1100$

(b) $1000 - 1000(0.25) = \$750$

(c) For the 20% increase: $1000 + 1000(0.20) = \$1200$. Then apply the 20% decrease: $1200 - 1200(0.20) = \$960$.

3. Without a calculator or computer, match each exponential formula to one of the graphs I-IV.

(a) $100(0.9)^t$ **(b)** $100(1.2)^t$ **(c)** $150(1.1)^t$ **(d)** $150(0.7)^t$

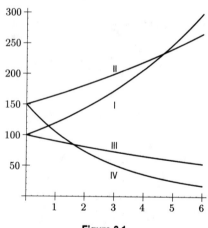

Figure 3.1

ANSWER:

(a) III

(b) I

(c) II

(d) IV

4. A population has size 600 at time $t = 0$, with t in years.

(a) If the population grows by 50 people per year, find a formula for the population, P, at time t.

(b) If the population grows by 8% per year, find a formula for the population, P, at time t.

ANSWER:

(a) $P = 600 + 50t$

(b) $P = 600(0.08)^t$

5. Table 3.1 contains values from an exponential or a linear function. Decide if the function is exponential or linear, find a possible formula for the function, and graph it.

Table 3.1

x	$f(x)$
0	1.8
1	1.5
2	1.2
3	0.9
4	0.6

ANSWER:
The rate of increase is the same for each interval: -0.3. Thus the function is linear, and its formula could be $f(x) = 1.8 - 0.3x$.

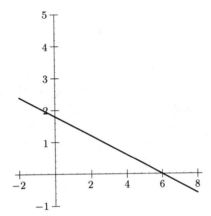

Figure 3.2

6. The table below shows v, the dollar value of a share of a certain stock, as a function of t, the time (in weeks) since the initial offering of the stock. Find a possible formula for $v(t)$.

Table 3.2

t	v
0	250.00
1	189.46
2	143.59
3	108.82
4	82.47
5	62.50
6	47.37

ANSWER:
By taking ratios of successive values of $v(t)$, we see that v decreases by the same factor 0.758, each week:

$$\frac{189.46}{250} \approx 0.758, \quad \frac{143.59}{189.46} \approx 0.758, \quad \frac{108.82}{143.59} \approx 0.758.$$

Thus, $v(t)$ seems to be an exponential function, whose formula is

$$v(t) = 250(0.758)^t.$$

7. In Figure 3.3, the functions $f, g, h,$ and p can all be written in the form $y = ab^t$. Which function has the largest value for a? Which has the largest value for b?

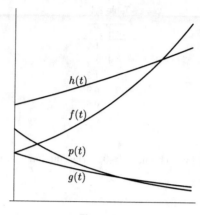

Figure 3.3

ANSWER:

$h; f$

8. Solve $y = 15(0.87)^x$ graphically for x if $y = 10$.

ANSWER:

$x = 2.9$

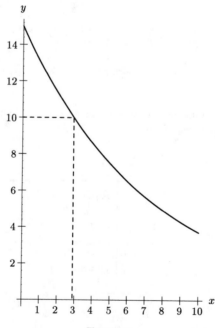

Figure 3.4

9. Without a calculator, match the functions $y = e^t$, $y = 2^t$, $y = e^{-t}$, and $y = 2^{-t}$ with the graphs in Figure 3.5

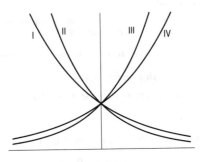

Figure 3.5

ANSWER:

$y = e^t$ is III, $y = 2^t$ is IV, $y = e^{-t}$ is II, and $y = 2^{-t}$ is I.

10. A population is 40,000 in year $t = 0$ and declines at a continuous rate of 6% per year.

    (a) Find a formula for $P(t)$, the population in year t.

    (b) By what percent does the population decrease each year? Why is this less than 6%?

    ANSWER:

(a) $P(t) = 40,000e^{-0.06t}$

(b) $e^{-0.06} = 0.942$, so the population decreases by $1 - 0.942 = 0.058$, or 5.8% per year. This is less than 6% because as members leave the population, the 6% decrease applies to a smaller population size.

11. What is the balance after 5 years if an account containing $800 earns 6% interest compounded

    (a) Annually

    (b) Monthly

    (c) Continuously?

    ANSWER:

(a)

$$B = 800 \cdot \left(1 + \frac{0.06}{1}\right)^5 = \$1070.58$$

(b)

$$B = 800 \cdot \left(1 + \frac{0.06}{12}\right)^{12(5)} = \$1079.08$$

(c)

$$B = 800e^{0.06(5)} = \$1079.89$$

Problems

12. A drug is injected into a patient's bloodstream over a five-minute interval. During this time, the quantity in the blood increases linearly. After five minutes the injection is discontinued, and the quantity then decays exponentially. Sketch a graph of the quantity versus time.

    ANSWER:

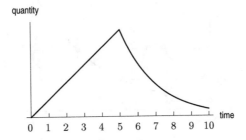

13. Sales of cassette tapes of music decreased by 6% per year over a period of 5 years. By what total percent did sales of cassette tapes decrease during this time period?

ANSWER:

Sales of cassettes decreased by a factor of 0.94 per year for 5 years. Thus, they decreased by a factor of

$$0.94^5 \approx 0.734,$$

which means sales dropped by 26.6% during this time period.

14. The population of Seattle has been growing at a rate of 5% per year. If the population was 100,000 in 1960, what was the projected population for 1998?

ANSWER:

We know that the equation for the population of Seattle must be in the form of

$$P(t) = P_0 \cdot k^t$$

where P_0 is the population at $t = 0$ and k is the factor by which the population changes every year. If we let 1960 be the year $t = 0$ we get

$$P(t) = 100{,}000k^t.$$

The fact that the population changes by 5% each year is equivalent to the statement that the population changes by a factor of 1.05 each year. Thus we get

$$P(t) = 100{,}000(1.05)^t.$$

The year 1998 is equivalent to $t = 38$ and we find that the projected population for the year 1998 is

$$P(38) = 100{,}000(1.05)^{38}$$
$$\approx 638{,}548.$$

15. In 1990, Peru's daily inflation rate was 1.2% per day. What was the corresponding annual inflation rate? Give your answer as a percent.

ANSWER:

If P_0 is the initial price and time t is in days, then $P(t) = P_0(1 + r)^t$. Taking $r = 0.012$, in one year the prices are given by

$$P(365) = P_0(1.012)^{365}.$$

If the yearly inflation rate is R, so that $P(365) = P_0(1 + R)$, we have

$$P_0(1 + R) = P_0(1.012)^{365}.$$

Thus,

$$1 + R = (1.012)^{365} = 77.7834$$
$$R = 76.7834 = 7678.34\%$$

16. The Internet is a network of computers allowing the transfer of information. Figure 3.6 gives a graph of N, the number of packets of information sent per month across the Internet (in billions), as a function of t, the number of years since 1990. Find a possible formula for $N(t)$.

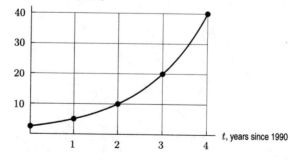

Figure 3.6

ANSWER:
Notice that N appears to double in size every year, from 2.5 to 5, to 10, to 20, and finally to 40. Thus, N appears to be an exponential function. The initial value is 2.5 and the value doubles every year, and so a possible formula is

$$N(t) = 2.5(2)^t.$$

17. The US population in 1996 was about 273.6 million. Assume that the population increases at the rate of 1.3% per year.

   (a) Write a formula for the population $P(t)$ of the US in millions, where t is the number of years since 1996.

   (b) According to your formula, what will be the population of the US in the year 2020? Round your answer to the nearest million (and, as always, include units).

   (c) Some demographers believe that the ideal population of the US would be about 130 million. According to your model, in what year did this occur?

   ANSWER:

   (a) $P(t) = 273.6 \cdot 1.013^t$.

   (b) $P(24) \approx 373$ million.

   (c) By finding the intersection of $y = 273.6 \cdot 1.013^t$ with $y = 130$ on the calculator, we obtain $t = -57.6$. Since $1996 - 57.6 = 1938.4$, the model predicts that it happened in 1938. (As an aside, the US population actually reached that size at around the year 1942).

18. The price of a certain item increases due to inflation. Let $p(t) = 7.50(1.058)^t$ give the price of the item as a function of time in years, with $t = 0$ in 1970.

   (a) What is the annual inflation rate?

   (b) What is the monthly inflation rate?

   (c) Describe as precisely as possible the meaning of $f^{-1}(20)$.

   (d) Approximate $f^{-1}(20)$ to two decimal places.

   ANSWER:

   (a) The annual inflation rate is 5.8%.

   (b) The monthly inflation rate is

   $$(1.058)^{\frac{1}{12}} - 1 \approx 0.471\%.$$

   (c) $f^{-1}(20) =$ the time at which the price reaches \$20.

   (d) Solve for t by using graphing or trial and error gives 17.40 years.

19. Kevin buys a new CD player for \$250.00, and finds two years later when he wants to sell it that it is only worth \$100.00. Find a formula for the value of the CD player if the value decreases:

   (a) linearly.

   (b) exponentially.

   ANSWER:

   (a) We know that when $t = 0$ the value of the CD player is \$250.00 and when $t = 2$ its value is \$100.00. If the function is linear
   $m = \frac{250-100}{0-2} = -75$ therefore $r(t) = -75t + 250$

   (b) If $r(t)$ is exponential $r(t) = ab^t$. Therefore when $t = 0$ we have $a = 250$ and $r(t) = 250b^t$. Since $r(2) = 100 = 250b^2$ we have $b = (0.4)^{\frac{1}{2}} = 0.63$. Therefore $r(t) = 250(0.63)^t$.

20. The graph of $P(t)$, an exponential function, follows.

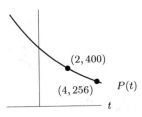

   (a) Find a formula for $P(t)$.

   (b) Suppose $P(t)$ represents a city's population, in thousands, t years after 1980. Evaluate the expression $P(10) - P(5)$. What does this expression represent in the context of the city's population?

ANSWER:

(a) Since $P(t)$ is an exponential function, we know that

$$400 = P(2) = a \cdot b^2 \quad \text{and}$$
$$256 = P(4) = a \cdot b^4.$$

From these two equations, we can calculate b,

$$\frac{256}{400} = \frac{ab^4}{ab^2} = b^2$$
$$b = \frac{16}{20} = \frac{4}{5}.$$

Using b, we can calculate a,

$$400 = a \left(\frac{4}{5}\right)^2$$
$$a = 400 \left(\frac{5}{4}\right)^2 = 625.$$

So,

$$P(t) = 625 \left(\frac{4}{5}\right)^t.$$

(b) Using our formula for $P(t)$,

$$P(10) - P(5) = 625 \left(\frac{4}{5}\right)^{10} - 625 \left(\frac{4}{5}\right)^5 = 625 \left(\frac{4}{5}\right)^5 \left(\left(\frac{4}{5}\right)^5 - 1\right) \approx -138.$$

The expressions $P(10) - P(5) \approx -138$ means that from the year 1985 to the year 1990 the city's population decreased by 138 (thousand).

21. A biologist measures the amount of contaminant in a lake four hours after a chemical spill and again 12 hours after the spill. She sets up two possible models to determine Q, the amount of chemical remaining in the lake as a function of t, the time in hours.

(a) In the first model, she assumes that the contaminant is leaving the lake at a constant rate. From her measurements, she then concludes that the contaminant is leaving at the rate of 5 tons/hour and that the lake will be free of contaminant 30 hours after the spill. Based on this model, find a formula for $Q(t)$, the quantity (in tons) of contaminant remaining in the lake as a function of time. What were the amounts measured at four and 12 hours?

(b) In her second model, she assumes that the amount of contaminant decreases exponentially. Using the data available, find another formula for $Q(t)$.

(c) She measures the contaminant again 19 hours after the spill and finds that 65 tons remain. Based on this evidence, which model seems best?

ANSWER:

(a) In this model we have $Q_1(t) = Q_0 - 5t$. In order for the contaminant to be gone after 30 hours, we require

$$Q_1(30) = 0 = Q_0 - 150$$
$$Q_0 = 150.$$

So

$$Q_1 = 150 - 5t \quad t \text{ in hours.}$$

(b) She measured the amount of contaminant at $t = 4$ hours and $t = 12$ hours. Thus, she must have used this information to contruct the first model. So, her measurements must have been $Q_1(4) = 150 - 5(4) = 130$ and $Q_1(12) = 150 - 5(12) = 90$. We will use $Q(4) = 130$ and $Q(12) = 90$ to construct the second model $Q_2(t) = a \cdot b^t$. First we solve for b:

$$\frac{Q_2(12)}{Q_2(4)} = \frac{a \cdot b^{12}}{a \cdot b^4} = \frac{90}{130}$$
$$b^8 = \frac{9}{13}$$
$$b = \left(\frac{9}{13}\right)^{1/8} \approx 0.9551.$$

Using b we can calculate a:

$$a \cdot b^4 = 130$$

$$a \cdot \left(\left(\frac{9}{13} \right)^{1/8} \right)^4 = 130$$

$$a \cdot \left(\frac{9}{13} \right)^{1/2} = 130$$

$$a = \frac{130}{\left(\frac{9}{13} \right)^{1/2}} \approx 156.24.$$

So

$$Q_2(t) = 156.24(0.9551)^t \quad t \text{ in hours.}$$

(c) For 19 hours after the spill, our two models predict:

$$Q_1(19) = 150 - 5(19) = 150 - 95 = 55$$
$$Q_2(19) = 156.24(0.9551)^{19} \approx 65.27.$$

So the second model, $Q_2(t) = 156.24(0.9551)^t$, seems best.

22. Let $f(x)$ be given by the table below:

x	$f(x)$
0	2
1	k
2	18

Find the value of k if:

(a) $f(x)$ is linear.

(b) $f(x)$ is exponential.

ANSWER:

(a) If $f(x)$ is linear $m = \frac{18-2}{2-0} = 8$ and $b = 2$. Therefore $f(x) = 8x + 2$ and $f(1) = k = 8(1) + 2 = 10$.

(b) If $f(x)$ is exponential $\frac{k}{2} = \frac{18}{k}$ and $k^2 = 36$. Therefore $k = 6$.

23. (a) Use your calculator to approximate any value(s) of x which satisfy $3 + x = 3(2)^x$.

(b) Describe the process you used to solve part (a) and explain why you think there are no additional solutions.

ANSWER:

(a) Graphically, the two functions look like the figure below. They intersect on the y-axis (at $x = 0, y = 3$) and once for $x < 0$. Solutions: $x = 0, x \approx -2.45$.

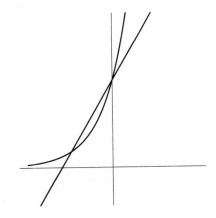

(b) Use the graph, trace and zoom features to graph $3 + x - 3(2)^x$ and see where this graph crossed the x-axis. There can be no more than two solutions, because after the exponential function overtakes the linear function it will always be above. The exponential function is never negative and thus will not cross the linear function as it becomes more and more negative.

24. The graph in Figure 3.7 shows two functions, one linear and one exponential. Find possible formulas for these two functions.

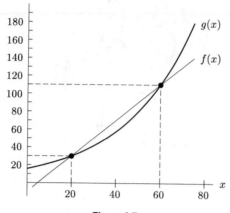

Figure 3.7

ANSWER:

The linear function is f. Therefore $f(x) = b + mx$ and $m = \frac{\Delta y}{\Delta x}$. Thus,

$$m = \frac{110 - 30}{60 - 20} = \frac{80}{40} = 2$$

Using the point $(x, y) = (20, 30)$ gives

$$30 = b + 2 \cdot 20$$
$$b = -10.$$

Thus, the linear function is given by $y = 2x - 10$.

The exponential function is g, and so $g(x) = ab^x$. Using the ratio method gives us

$$\frac{g(60)}{g(20)} = \frac{ab^{60}}{ab^{20}} = \frac{110}{30}$$
$$b^{40} = \frac{11}{3}$$
$$b = \left(\frac{11}{3}\right)^{1/40} \approx 1.033.$$

Therefore, using $(x, y) = (20, 30)$, we have

$$g(20) = ab^{20} = a(1.033)^{20} = 30$$
$$a = \frac{30}{(1.033)^{20}} \approx 15.67.$$

Thus, $g(x) = 15.67(1.033)^x$.

25. Each of the functions in the table is increasing, but each increases in a different way. One is linear, one is exponential, and one is neither. Match the graphs to the data in the table, and find a formula for the linear and exponential functions. For the remaining function, write "neither".

Table 3.3

t	g(t)	h(t)	k(t)
1	22.40	10	2.2
2	25.09	20	2.5
3	28.10	29	2.8
4	31.47	37	3.1
5	35.25	44	3.4
6	39.48	50	3.7

(a)

(b)

(c)

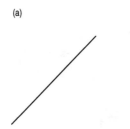  

Figure 3.8

ANSWER:

(a) Figure (a): $k(t)$. Linear. Formula is $k(t) = 0.3t + 1.9$.
(b) Figure (b): $h(t)$. Neither linear nor exponential.
(c) Figure (c): $g(t)$. Exponential. Formula is $g(t) = 20(1.12)^t$.

26. Suppose the amount of ozone in the atmosphere is decreasing exponentially at a continuous rate of 0.25% per year. How many years will it take before half the ozone has disappeared?

ANSWER:

If Q_0 is the current amount of ozone in the air and Q is the amount at time t, we are given

$$Q = Q_0 e^{-0.0025t}.$$

We want to find t when $Q = Q_0/2$, so

$$\frac{Q_0}{2} = Q_0 e^{-0.0025t}$$
$$-\ln 2 = -0.0025t$$
$$t = \frac{\ln 2}{0.0025} = 277 \text{ years.}$$

27. Figure 3.9 gives a graph of $C = f(t)$, where C is the computer hard disk drive capacity (in hundreds of megabytes) that can be bought for $500 and where t is the number of years since 1989.

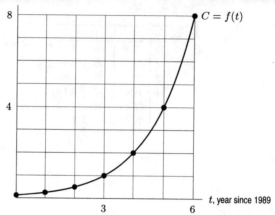

C, capacity (in 100s of megabytes)

Figure 3.9

(a) If $f(t) = a \cdot e^{kt}$, find a and k.

(b) Assuming that the trend displayed by the graph continued, estimate the value of $f(10)$. Explain the significance of your answer in terms of hard drives. Be specific.

(c) Assuming that the trend displayed by the graph continued, estimate the value of $f^{-1}(25)$. Explain the significance of your answer in terms of hard drives. Be specific.

ANSWER:

(a) We assume

$$f(t) = a \cdot e^{kt}.$$

Substitute $(3, 1)$ and $(6, 8)$

$$1 = a \cdot e^{3k}$$
$$8 = a \cdot e^{6k}$$

Dividing the equations gives

$$\frac{8}{1} = \frac{e^{6k}}{e^{3k}}$$

So,

$$e^{6k-3k} = 8$$
$$e^{3k} = 8$$

Take the natural log of both sides:

$$\ln e^{3k} = \ln 8$$

So,

$$3k = \ln 8$$
$$k = \frac{\ln 8}{3}$$
$$\approx 0.6931$$

Then

$$a = \frac{1}{e^{3k}} = 0.125$$

so

$$f(t) = 0.125 \cdot e^{0.6931t}$$

(b)

$$f(10) = 0.125 \cdot e^{(0.6931)(10)} = 128$$

This means that in year $t = 10$ (that is, in 1999) assuming the trend continued, $500 purchased a hard drive with a 12,800 megabyte capacity.

(c) Since $C = f(t)$ gives the capacity in years t, the quantity $f^{-1}(25)$ gives the *year* in which the capacity reached $C = 2500$ megabytes. This means we must solve $f(t) = 25$ for t.

$$0.125e^{0.6931t} = 25$$
$$e^{0.6931t} = 200$$
$$0.6931t = \ln 200$$
$$t = \frac{\ln 200}{0.6931}$$
$$\approx 7.6$$

$$f^{-1}(25) = 7.6$$

The significance of this is that, assuming the trend continued, it took 7.6 years, or well into 1996, for $500 of hard drive capacity to reach 2500 megabytes.

28. The price of a certain item is increasing due to inflation. If $p = f(t)$ gives the price of the good t years after 1970, then $f(t)$ is well approximated by the formula

$$f(t) = 7.50(1.058)^t.$$

   (a) At what continuous annual rate is the price increasing?
   (b) By what percent does the price increase each month?
   (c) Describe the economic significance of the fact that $f^{-1}(20) \approx 17.40$.

   ANSWER:

   (a) $f(t) = 7.50(1.058)^t = 7.50(e^{\ln 1.058})^t \approx 7.50e^{0.0564t}$ so 5.64% is continuous annual growth rate.
   (b) $f(1/12) = 7.50(1.058)^{1/12} = 7.50(1.00471)$. So, prices rise by 0.471% per month.
   (c) $f^{-1}(20) \approx 17.40$ tells us that the price of the item rises to $20 after 17.4 years, or by May of 1987.

29. Find formulas for the functions representing the growth of each of the following quantities.

   (a) This quantity begins at N in year $t = 0$ and grows at a constant percent annual rate of $r\%$.
   (b) This quantity begins at N in year $t = 0$ and grows at a continuous percent annual rate of $r\%$.
   (c) This quantity grows in the same way as the one in part (b) except that it starts out twice as large.
   (d) This quantity grows in the same way as the one in part (b) except that it starts out 30% larger.
   (e) This quantity grows in the same way as the one in part (b) except that it grows at a continuous annual rate that is half as fast.

   ANSWER:

   (a) $f(t) = N(1 + \frac{r}{100})^t$
   (b) $f(t) = Ne^{\frac{rt}{100}}$
   (c) $f(t) = 2Ne^{\frac{rt}{100}}$
   (d) $f(t) = 1.3Ne^{\frac{rt}{100}}$
   (e) $f(t) = Ne^{\frac{rt}{200}}$

30. Let $P(t) = 1000e^{0.041t}$ give the size of a population of animals in year t.

   (a) Briefly describe this population in words. Be specific.
   (b) Evaluate $P(15)$. Explain what this tells you about the population.
   (c) Solve $P(t) = 5000$ for t. What do your solution(s) tell you about the population?

   ANSWER:

   (a) This population begins with 1000 members and grows at a continuous annual rate of 4.1%.
   (b) We have (using a calculator) $P(15) = 1000e^{0.041(15)} = 1849.66 \approx 1850$. This tells us that after 15 years the population will have 1850 members.
   (c) We have

$$1000e^{0.041t} = 5000$$
$$e^{0.041t} = 5$$

$$\ln(e^{0.041t}) = \ln 5$$
$$0.041t = \ln 5$$
$$t = \frac{\ln 5}{0.041} = 39.25.$$

This tells that the population will reach 5000 members in just under 40 years.

31. (a) Make a table of values for $f(x) = (0.7)^x$ for $x = -3, -2, -1, 0, 1, 2, 3$.
(b) Graph $f(x)$. Describe the graph in words.
ANSWER:

(a)

Table 3.4

x	$f(x)$
-3	2.92
-2	2.04
-1	1.43
0	1
1	0.7
2	0.49
3	0.34

(b)

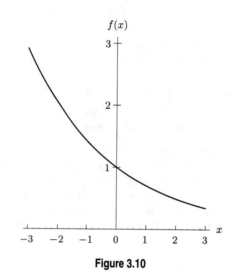

Figure 3.10

The graph is concave up and crosses the y−axis at 1. As $x \to -\infty$, $f(x) \to \infty$. As $x \to +\infty$, $f(x) \to 0$.

32. The population of a city is increasing exponentially. In 1998 the city had a population of 30,000. In 2002, the population was 48,000.

(a) Give a formula for the population P, of the city as a function of time, t, in years since 1998.
(b) Use a graph to estimate when the population will be 100,000.
ANSWER:

(a) We know that $P = ab^t$, and $a = 30,000$, so we need to solve for b. At $t = 4$ (in 2002)

$$48,000 = 30,000b^4$$

$$1.6 = b^4$$

$$b = 1.125.$$

Thus, $P = 30,000(1.125)^t$.

(b)

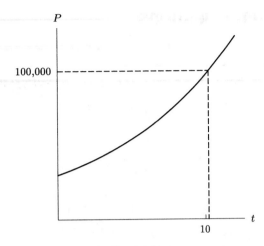

Figure 3.11

Thus, $P = 100,000$ just after $t = 10$, or sometime in the year 2008.

Chapter 4 Exam Questions

Exercises

1. Rewrite the statements using exponents instead of logs.

(a) $\ln\left(\dfrac{1}{2}\right) = -0.693$

(b) $\log r = s$

ANSWER:

(a) $e^{-0.693} = \dfrac{1}{2}$

(b) $10^s = r$

2. Rewrite the statements using logs.

(a) $10^{-5} = 0.00001$

(b) $e^{2a} = b$

ANSWER:

(a) $\log(0.00001) = -5$

(b) $\ln b = 2a$

3. Solve the equations using logs.

(a) $3^x = 8$

(b) $\dfrac{1}{2} = \left(\dfrac{1}{3}\right)^x$

ANSWER:

(a)

$$3^x = 8$$
$$\log 3^x = \log 8$$
$$x \log 3 = \log 8$$
$$x = \frac{\log 8}{\log 3} = 1.893$$

(b)

$$\frac{1}{2} = \left(\frac{1}{3}\right)^x$$
$$\ln\left(\frac{1}{2}\right) = \ln\left(\frac{1}{3}\right)^x$$
$$\ln\left(\frac{1}{2}\right) = x \ln\left(\frac{1}{3}\right)$$
$$x = \frac{\ln\left(\frac{1}{2}\right)}{\ln\left(\frac{1}{3}\right)} = 0.631$$

4. Find the doubling time for a bank account that is growing by 4.6% per year.
ANSWER:
If P_0 is the initial balance, then

$$2P_0 = P_0 e^{0.046t}$$
$$2 = e^{0.046t}$$
$$\ln 2 = 0.046t$$
$$t = \frac{\ln 2}{0.046} = 15.07.$$

Thus, the doubling time is about 15 years.

5. Convert $Q = \dfrac{1}{5}e^{0.06t}$ to the form $Q = ab^t$.
ANSWER:
$$Q = \frac{1}{5}e^{0.06t} = \frac{1}{5}\left(e^{0.06}\right)^t = \frac{1}{5}(1.06)^t.$$

6. Write the exponential function $y = 20e^{-0.04t}$ in the form $y = ab^t$. Find b accurate to four decimal places. If t is measured in years, give the percent annual growth or decay rate and the continuous percent growth or decay rate per year.

   ANSWER:

   $y = 20e^{-0.04t} = 20 \left(e^{-0.04}\right)^t = 20(0.9608)^t$. The continuous percent decay rate is 4%, and the percent annual decay rate is $1 - 0.9608 = 0.0392 = 3.92\%$.

7. Without a calculator, match the functions $y = e^{-x}$, $y = 10^{-x}$, $y = \ln{(-x)}$, and $y = \log{(-x)}$ with the graphs in Figure 4.1

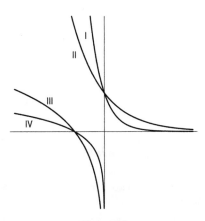

Figure 4.1

   ANSWER:

   $y = e^{-x}$ is II; $y = 10^{-x}$ is I; $y = \ln{(-x)}$ is III; and $y = \log{(-x)}$ is IV.

8. Graph $y = e^{x-1}$. Label all asymptotes and intercepts.

   ANSWER:

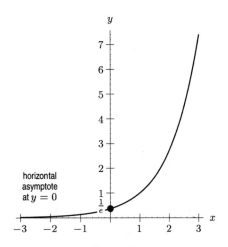

Figure 4.2

9. Graph $y = 1 + \ln{(1 - x)}$. Label all asymptotes and intercepts. State the domain of the function.

ANSWER:

The domain is where $\ln(1-x)$ is defined, which is where $1-x > 0$. Thus, the domain is $(-\infty, 1)$.

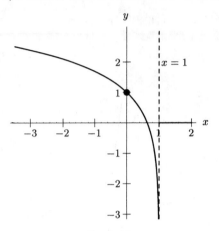

Figure 4.3

10. You wish to graph the following quantities on a standard sheet of paper. On which should you use a linear scale? On which should you use a logarithmic scale? Why?

   (a) The distance from the earth (in millions of kilometers) of each of the planets in our solar system.
   (b) The number of moons revolving around each of the planets in our solar system.

   ANSWER:

   (a) Logarithmic scale because of the wide range of large values involved
   (b) Linear scale because the values will not vary as much

11. On the log scale in Figure 4.4, mark 5, 10, and 50.

Figure 4.4

ANSWER:

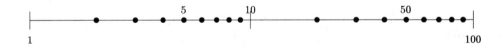

Figure 4.5

12. The following table gives z as a function of x,

x	3	4	5	6
z	8	11	14	17

   (a) Find a formula for z in terms of x.
   (b) If $z = \ln y$, find a formula for y in terms of x.

ANSWER:

(a) Notice that z goes up 3 as x goes up 1 thus the slope of our function will be 3. Also, $z(0) = -1$ so the desired formula is $z = 3x - 1$.

(b)

$$z = \ln y = 3x - 1$$
$$e^{\ln y} = e^{3x-1}$$
$$\text{Thus } y = e^{3x-1}$$

13. The following table gives v as a function of w,

w	2	4	6	8
v	-4	-8	-12	-16

(a) Find a formula for v in terms of w.

(b) If $w = \ln x$, and $v = \ln y$, find a formula for y in terms of x.

ANSWER:

(a) $v = -2w$.

(b)

$$v = -2w$$
$$\ln y = -2\ln x$$
$$e^{\ln y} = e^{-2\ln x}$$
$$= e^{\ln x^{-2}}$$
$$y = x^{-2}$$

Problems

14. Let $f(x) = \log(9x)$, and $g(x) = \log(3x)$, and $h(x) = \dfrac{f(x) - g(x)}{f(x) + g(x)}$, with $x > 0$. Find and simplify a formula for $h(x)$ in terms of x.

ANSWER:
$$h(x) = \frac{\log 9x - \log 3x}{\log 9x + \log 3x} = \frac{\log(\frac{9x}{3x})}{\log(9x \cdot 3x)} = \frac{\log 3}{\log(27x^2)}.$$

15. The notation $\log_b x$ means that the log base is unspecified. Given the following: $\log_b x = 2.1$, $\log_b y = 3.4$, and $\log_b z = 4.5$, use the properties of logs to find $\log_b(\dfrac{xy^2}{z})$. (Show your work.)

ANSWER:

$$\log_b\left(\frac{xy^2}{z}\right) = \log_b(xy^2) - \log_b(z)$$
$$= \log_b x + 2\log_b y - \log_b z$$
$$= 2.1 + 2(3.4) - 4.5 = 2.1 + 6.8 - 4.5 = 8.9 - 4.5 = 4.4$$

16. Simplify the following completely:

(a) $\ln\left(x^2 e^{-\ln\sqrt{x}}\right)$

(b) $\dfrac{\log A^3 - \log\sqrt{A}}{\log(e^{\ln e})}$

ANSWER:

(a)

$$\ln(x^2 e^{-\ln\sqrt{x}}) = \ln x^2 + \ln(e^{-\ln\sqrt{x}})$$
$$= 2\ln x + (-\ln\sqrt{x})$$
$$= 2\ln x - \frac{1}{2}\ln x$$
$$= \frac{3}{2}\ln x$$

(b)

$$\frac{\log A^3 - \log \sqrt{A}}{\log(e^{\ln e})} = \frac{\log \frac{A^3}{\sqrt{A}}}{\log e}$$

$$= \frac{\log A^{\frac{5}{2}}}{\log e}$$

$$= \frac{5}{2}\frac{\log A}{\log e}$$

Some advanced students may recognize this as $\frac{5}{2}\ln A$.

17. Let $n = \log p$, and $m = \log q$. If possible, rewrite the following expressions in terms of n and m and simplify.

   (a) $\log(pq)$
   (b) $\log \sqrt{p}$
   (c) $\log(0.0001q)$
   (d) $\log \dfrac{p^3}{q^5}$

   ANSWER:

   (a) $\log pq = \log p + \log q = n + m$
   (b) $\log \sqrt{p} = \log p^{\frac{1}{2}} = \frac{1}{2}\log p = \frac{1}{2}n$
   (c) $\log(0.0001q) = \log(0.0001) + \log q = \log 10^{-4} + \log q = -4 + \log q = -4 + m$
   (d) $\log \dfrac{p^3}{q^5} = \log p^3 - \log q^5 = 3\log p - 5\log q = 3n - 5m$

18. Solve the following equations for t exactly.

   (a) $\log(t - 175) = 3$
   (b) $200e^{0.6t} = 300e^{0.8t}$

   ANSWER:

   (a)

   $$\log(t - 175) = 3$$
   $$10^{\log(t-175)} = 10^3$$
   $$t - 175 = 1000$$
   $$t = 1175$$

   (b)

   $$200e^{0.6t} = 300e^{0.8t}$$
   $$\frac{e^{0.6t}}{e^{0.8t}} = \frac{300}{200}$$
   $$e^{-0.2t} = \frac{3}{2} \qquad \text{Using an exponent rule}$$
   $$-0.2t = \ln 32$$
   $$t = \frac{\ln \frac{3}{2}}{-0.2} = -5\ln \frac{3}{2}$$

19. In September 1994 the United Nations hosted the Conference on Population and Development in Cairo, Egypt. According to *The Sciences* magazine, the world population of 2.5 billion people in 1950 had grown to 5.5 billion by the 1990 census.

   (a) Determine an exponential function to model the world population, P, as a function of time, t, in years since 1950 (i.e., let $t = 0$ represent the year 1950.) [Show your base to four decimal places.]
   (b) According to this model, in what year will the world population reach eleven billion people?
   (c) Recently, the rate of growth of the world's population has slowed to about 100 million people per year. Predictions are that the world population will continue to increase at 100 million people per year for some time. Using this information and letting $t = 0$ in 1990 when the population was 5.5 billion, determine a new model for the world population, P, as a function of time, t, in years since 1990. [Note: 100 million = 0.1 billion.]
   (d) According to the model from part (c), in what year will the world population reach eleven billion people?

ANSWER:

(a) We know the model is of the form $P = P_0 b^t$. Given $(0, 2.5)$ and $(40, 5.5)$, we get $P_0 = 2.5$ and $5.5 = 2.5(b)^{40}$, thus, $b = \left(\frac{5.5}{2.5}\right)^{\frac{1}{40}} = 1.0199$. Therefore

$$P = 2.5(1.0199)^t.$$

(b) We want t so that

$$11 = 2.5(1.0199)^t.$$

One way to solve for t is to graph both sides and trace to the intersection point. We find $t \approx 75.19$. Thus, the population reaches 11 billion in 2025. Alternatively, we can use logs:

$$11 = 2.5(1.0199)^t$$
$$\frac{11}{2.5} = 1.0199^t$$
$$\log \frac{11}{2.5} = \log 1.0199^t$$
$$\log \frac{11}{2.5} = t \cdot \log 1.0199$$
$$t = \frac{\log(11/2.5)}{\log 1.0199} \approx 75.19.$$

(c) If population is increasing by 100 million per year, then the increase is *linear*. If we want years since 1990, then $b = 5.5$ and $m = 0.1$ in $y = b + mx$. Thus, $P = 5.5 + 0.1t$ reflects this trend.

(d) We want t so that

$$11 = 5.5 + 0.1t$$
$$5.5 = 0.1t$$
$$t = 55 \quad \text{so} \quad 1990 + 55 = 2045.$$

This model predicts that the world population will reach 11 billion in 2045.

20. A population of bacteria quadruples every 3 minutes.

   (a) By what percent does the population increase each minute?
   (b) How many minutes does it take for the population to double?
   (c) If $P(t)$ is the population at time t, and if n is the number of bacteria at time t_0, find $P(t_0 + 6)$ in terms of n, and explain the significance of your answer in terms of the bacteria.
   (d) Given the definition of n from part (c), evaluate $P^{-1}(2n)$ in terms of t_0, and explain the significance of your result in terms of the bacteria.

   ANSWER:

   Since we know that the population quadruples in three minutes, we know that the function will be of the form

$$P(t) = P_0(4)^{t/3}.$$

   (a) Looking at the formula for $P(t)$ we see that in one minute the population increases by

$$(4^{1/3} - 1)100 = (\sqrt[3]{4} - 1)100 \text{ percent} \approx 58.74\%$$

   (b) Since the population quadruples every 3 minutes, it must double every

$$1.5 \text{ minutes.}$$

   (c) Since we know that in three minutes the population quadruples, we know that 6 minutes from time t_0 the population will be sixteen times as large or

$$P(t_0 + 6) = 16n.$$

   (d) We are looking for the time when the population is twice as large as it was at time t_0. Since we know that the population doubles every 1.5 minutes, this means that

$$P^{-1}(2n) = t_0 + 1.5.$$

21. A radioactive isotope decays according to the formula $N(t) = 40e^{-0.02t}$, where $N(t)$ gives the amount still present after t hours. Find the half-life of the isotope.

ANSWER:

Let $t_{1/2}$ denote the half-life of the decay. Then $t_{1/2}$ hours after the start, the amount of isotope remaining will be half of the amount initially present. So

$$N(t_{1/2}) = 20 = 40e^{-0.02t_{1/2}}$$
$$e^{-0.02t_{1/2}} = 0.5$$

Taking natural logs

$$\ln e^{-0.02t_{1/2}} = \ln 0.5$$
$$-0.02t_{1/2} = -0.6931$$
$$t_{1/2} = 34.66 \text{ hours}$$

22. The formulas below give five different populations as functions of time, where t is in years.

$$P_1 = 15000 + 50t$$
$$P_2 = 5000(0.80)^t$$
$$P_3 = 5000e^{-0.22t}$$
$$P_4 = 8000(2)^{t/6}$$
$$P_5 = 8000(2)^{6t}$$

Match each of the following statements with one of the formulas above.

(a) This population's annual percent rate of decrease is the greatest of the populations given.
(b) This population is increasing at a decreasing percent rate.
(c) This population doubles in size in 6 years.
(d) This population doubles in size every 2 months.

ANSWER:

(a) Only populations P_2 and P_3 are decreasing. P_2 is decreasing by 20% per year. P_3 is decreasing at the continuous rate of 22% per year and since $P_3 = 5000(e^{-0.22})^t \approx 5000(0.8025)^t$, this amounts to an annual decrease of $\approx 19.75\%$. Thus the population with the greatest rate of decrease is P_2.

(b) All the populations except P_1 grow exponentially. Thus P_1 is the only population that does not grow at a constant percent rate. Since it grows by 50 people per year, this increase represents a smaller percentage of the population over time. So the answer is P_1.

(c) P_4
(d) P_5

23. Suppose that $Q = A(1 + r)^{t/k}$, where k is a positive constant and t is in years.

(a) Suppose $k = 12$ and $r = 0.05$. What does this tell you about the growth rate of this function?
(b) Suppose $k = 20$ and $r = 1$. What does this tell you about the growth rate of this function? What is another common name for the value of k in this example?
(c) Suppose $k = 2000$ and $r = -0.5$. What does this tell you about the growth rate of this function? What is another common name for the value of k in this example?

ANSWER:

(a) For $k = 12$ and $r = 0.05$, $Q = A(1.05)^{t/12}$. This function describes a quantity growing at a rate of 5% every 12 years.
(b) For $k = 20$, and $r = 1$, $Q = A(2)^{t/20}$. This function describes a quantity that doubles every 20 years. (k is the doubling time.)
(c) For $k = 2,000$, and $r = -0.5$, $Q = A(0.5)^{t/2,000}$. This function describes a quantity with a half-life of 2,000 years. (k is the half-life.)

24. The acidity of a solution is measured by its pH. If [H^+] represents the concentration of hydrogen ions (in moles/liter) in the solution, the pH is defined by

$$pH = -\log[H^+].$$

A solution with pH<7 is said to be an acid; a solution with pH>7 is said to be a base.

(a) If $\left[H^+\right] = 0.00008$ for beer, is beer an acid or a base?

(b) Solve for [H^+] in terms of pH.

(c) What is the concentration of hydrogen ions if the solution has pH $= 2.7$?

ANSWER:

(a) For beer, pH $= -\log(0.00008) = 4.1$, so beer is an acid.

(b) Since $-\log[H^+] = pH$, we have

$$\log[H^+] = -pH.$$

So we have

$$\left[H^+\right] = 10^{-pH} = \frac{1}{10^{pH}}.$$

(c) If pH $= 2.7$, then

$$[H^+] = \frac{1}{10^{2.7}} = 0.001995 \text{ moles/liter.}$$

25. A quantity which is decaying exponentially has a half-life of 3 weeks. The time constant is defined to be the amount of time required for the quantity to decrease to $1/e$ of the original quantity.

(a) Without doing any calculations, decide if the time constant is larger or smaller than the half-life. Give a reason for your answer.

(b) Calculate the time constant. Give an exact answer and an answer accurate to one decimal place.

ANSWER:

(a) Since the half-life is the amount of time required for the quantity to decrease to $1/2$ of the original quantity and since $1/2 > 1/e$ (because $e > 2$), the time constant will be larger than the half-life (because it takes more time to decay further).

(b) Because the quantity decays exponentially, we have

$$Q = Q_0 e^{-kt}.$$

In addition, since $t = 3$ when $Q = Q_0/2$, we have

$$\frac{1}{2} = e^{-k(3)}$$

so

$$k = -\frac{\ln(1/2)}{3} = \frac{\ln 2}{3}.$$

We want to find the value of t making $Q = Q_0/e$, or

$$\frac{1}{e} = e^{-kt} = e^{-(\ln 2)t/3}.$$

Since $1/e = e^{-1}$, we have

$$e^{-1} = e^{-(\ln 2)t/3}$$

$$-1 = -\frac{(\ln 2)t}{3}$$

$$t = \frac{3}{(\ln 2)} \approx 4.3 \text{ weeks.}$$

26. Suppose y is given as follows:

$$y = t \cdot 2^x + 3$$

(a) Graph y against t for a positive constant x. Label all intercepts and asymptotes. (Your answers might contain x.)

(b) Graph y against x for a positive constant t. Label all intercepts and asymptotes. (Your answers might contain t.)

ANSWER:

(a) If x is constant, 2^x is the slope of this linear function. The y-intercept is 3. When $y = 0$, we have $t = -3/2^x$. So the t-intercept is at $-3/2^x$. See Figure 4.6.

(b) If t is a constant, the function is exponential, with a horizontal asymptote of 3. Thus the y-intercept is $t \cdot 2^0 + 3 = t + 3$ and the horizontal asymptote of $y = 3$. See Figure 4.7.

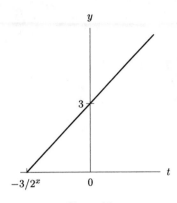

Figure 4.6

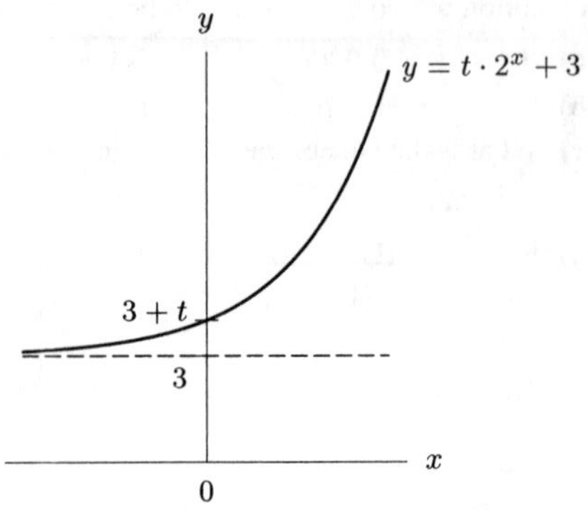

Figure 4.7

27. Figure 4.8 gives a graph of $y = \ln x$. The quantities y_1, y_2, and n represent fixed constants.

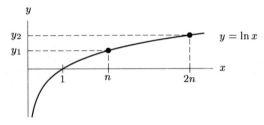

Figure 4.8

(a) Evaluate $y_2 - y_1$, and simplify your answer.

(b) Evaluate $e^{y_2} - 2e^{y_1}$, and simplify your answer.

(c) Evaluate $\ln(1/n)$. Your answer may include the constants y_1 and/or y_2. Plot the point $(1/n, \ln(1/n))$ on the graph.

ANSWER:

From graph, $y_1 = \ln n$ and $y_2 = \ln 2n$.

(a)

$$y_2 - y_1 = \ln 2n - \ln n$$
$$= \ln \frac{2n}{n} \qquad \text{using a log property}$$
$$= \ln 2.$$

(b)

$$e^{y_2} - 2e^{y_1} = e^{\ln 2n} - 2e^{\ln n}$$
$$= 2n - 2(n)$$
$$= 0.$$

(c) Since

$$\ln(1/n) = \ln(n^{-1}) = -\ln n,$$

we have

$$\ln(1/n) = -y_1.$$

This value is indicated on Figure 4.9

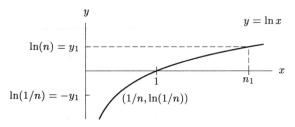

Figure 4.9

28. Find the equation of the line L in Figure 4.10. (Your answer will contain b)

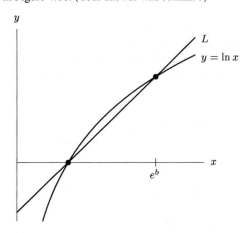

Figure 4.10

ANSWER:

Since $(1, 0)$ is the x-intercept of $y = \ln x$, and since $\ln(e^b) = b$, we know two points on the line are $(1, 0)$ and (e^b, b). So the equation of L is

$$y - 0 = m(x - 1).$$

The slope of L is

$$m = \frac{b - 0}{e^b - 1},$$

so we have

$$y = \left(\frac{b}{e^b - 1}\right)(x - 1).$$

29. Suppose $f(x) = \log x$. Are the following statements true or false?

(a) The graph of $f(x)$ goes through the point $(0, 1)$.
(b) The graph of $f(x)$ goes through the point $(10, 1)$.
(c) The inverse of $f(x)$ is the function $f^{-1}(x) = 1/\log x$.
(d) The range of $f(x)$ is all positive numbers.

ANSWER:

(a) False, since if $x = 0$, $\log x$ is not defined.
(b) True, since $\log 10 = 1$.
(c) False, this is the reciprocal. The inverse is $f^{-1}(x) = 10^x$.
(d) False, the range is all real numbers.

30. True or false?

(a) If $f(x) = 3^{-x}$, then $f(x)$ is an exponential function.
(b) If $2^x = 3$, then $x = \log 3 - \log 2$.
(c) If $x < 0$, then $4^x < 0$.
(d) If $x < 0$, then $(0.75)^x < (0.5)^x$.
(e) If $0 < n < m$, then $\log\left(\frac{1}{n}\right) < \log\left(\frac{1}{m}\right)$.

ANSWER:

(a) True; $f(x) = 3^{-x} = (1/3)^x$.
(b) False; $x = (\log 3)/(\log 2)$.
(c) False; $4^x > 0$ for all x.
(d) True; if $x = -1$, then $(0.75)^{-1} = 4/3$ and $(0.5)^{-1} = 2$.
(e) False; $\frac{1}{n} > \frac{1}{m}$ so $\log\left(\frac{1}{n}\right) > \log\left(\frac{1}{m}\right)$.

31. Suppose the population of a certain city is distributed in suburbs around the city's center. Let P be the number of people (in millions) living within r miles of the city's center. Then P can be approximated by the formula

$$P = \frac{4}{1 + 500e^{-2r}}.$$

Based on a graph of P, answer the following questions.

(a) Including suburbs, about how many people live in the city?
(b) Including suburbs, about how far is it from the city's center to the edge of town?
(c) About how far from the city's center are the most heavily populated neighborhoods?

ANSWER:

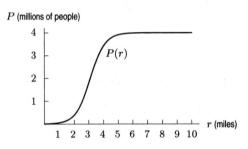

Figure 4.11

(a) About 4 million
(b) About 7 miles (answers may vary from 6 to 10)
(c) 3-4 miles away

32. Suppose you want to plot the daily low temperature recorded in Anchorage, Alaska for one year. On the horizontal axis you plot the day (1 to 365) and on the vertical axis the low temperature in degrees Celsius. Explain why it is possible, or not possible, to use a log scale on the vertical axis.

ANSWER:

Anchorage temperatures are both positive and negative. However, data that includes both positive and negative values can not be represented on a log scale. Thus, a log scale cannot be used here.

33. On the log scale in Figure 4.12, a point is marked halfway between the two endpoints. To two decimal places, what number does this midpoint represent?

10^0 10^1

Figure 4.12

ANSWER:
On a log scale this midpoint represents $10^{0.5} \approx 3.16$.

34. Given the graph in Figure 4.13, find a formula for y in terms of x.

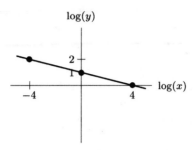

Figure 4.13

ANSWER:
The graph is linear with a slope of $-1/4$ and a vertical intercept of 1. Thus,

$$\log(y) = -\frac{1}{4}\log(x) + 1.$$

Now we solve for y in terms of x:

$$10^{\log y} = 10^{-\frac{1}{4}\log x + 1}$$
$$y = \left(10^{-\frac{1}{4}\log x}\right)\left(10^1\right)$$
$$y = 10\left(10^{\log x}\right)^{-1/4}$$
$$y = 10x^{-1/4}$$

35. Draw a line segment about 4 inches long. On it, choose an appropriate linear scale and mark points that represent 2^0, 2^{-1}, 2^{-2}, 2^{-3}, 2^{-4}, and 2^{-5}. Repeat the process with a logarithmic scale. Compare the locations of the points on the two scales.
ANSWER:

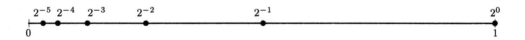

Figure 4.14

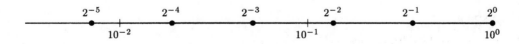

Figure 4.15

The points bunch up near 0 on the linear scale. They are more evenly spaced on the logarithmic scale.

Chapter 5 Exam Questions

Exercises

1. Complete the following table using $f(x) = x^2 + x$, $g(x) = f(x-1)$, and $h(x) = f(x) - 1$. Graph the three functions, and explain how the graphs of g and h are related to the graph of f.

Table 5.1

x	-3	-2	-1	0	1	2	3
f (x)							
g (x)							
h (x)							

ANSWER:

Table 5.2

x	-3	-2	-1	0	1	2	3
f (x)	6	2	0	0	2	6	12
g (x)	12	6	2	0	0	2	6
h (x)	5	1	-1	-1	1	5	11

g is one unit right, h is one unit down.

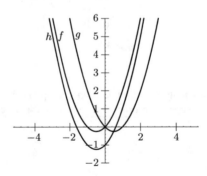

Figure 5.1

2. Write a formula and graph the transformations of $r(t) = e^t$ for $t > 0$.
   (a) $s_1 = r(t - 2)$
   (b) $s_2 = r(t) - 2$

   ANSWER:
   (a) $s_1 = e^{t-2}$. See Figure 5.2.
   (b) $s_2 = e^t - 2$. See Figure 5.3.

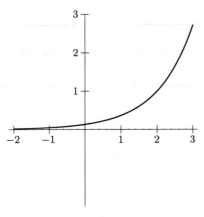

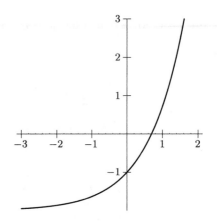

Figure 5.2   Figure 5.3

3. Let $f(x) = \ln x$, $g(x) = \ln(x+2)$, and $h(x) = \ln x + 2$ for $x > 0$. How do the graphs of $g(x)$ and $h(x)$ compare to the graph of $f(x)$?

ANSWER:

The graph of $g(x)$ is two units to the left, the graph of $h(x)$ is 2 units up.

4. Explain in words the effect of the transformation $f(x + 2a) - b$ on the graph of the function $f(x)$. Assume a and b are positive constants.

ANSWER:

Shift left $2a$, then down b.

5. Graph $y = g(x) = e^x$ and $y = g(-x)$ on the same set of axes. How are these graphs related? Give an explicit formula for $y = g(-x)$.

ANSWER:

The graph of $g(-x)$ is a reflection of the graph of $g(x)$ across the y−axis.

$$y = e^{-x}$$

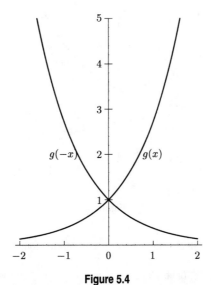

Figure 5.4

6. Give a formula and graph for each of the transformations of $m(n) = n^2 - n + 3$.

(a) $y = -m(n)$ (b) $y = m(-n)$

ANSWER:

(a) $y = -(n^2 - n + 3) = -n^2 + n - 3$.

(b) $y = (-n)^2 - (-n) + 3 = n^2 + n + 3$.

(a)

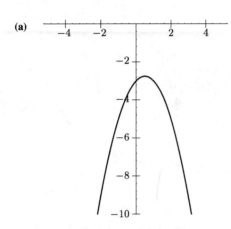

Figure 5.5

(b)

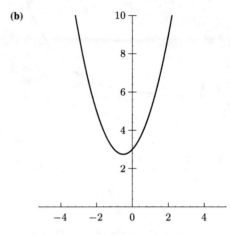

Figure 5.6

7. Show that the function is even, odd, or neither.

(a) $f(x) = x^4 - x^2 + 7$ (b) $f(x) = x^3 + 2x^2 + 4$

ANSWER:

(a) $f(-x) = (-x)^4 - (-x)^2 + 7 = x^4 - x^2 + 7 = f(x)$, so $f(x)$ is even.

(b) $f(-x) = (-x)^3 + 2(-x)^2 + 4 = -x^3 + 2x^2 + 4 \neq f(x)$, so $f(x)$ is not even. Also, $-f(x) = -(x^3 + 2x^2 + 4) = -x^3 - 2x^2 - 4 \neq f(-x)$, so $f(x)$ is not odd either.

8. Graph and label $f(x)$, $2f(x)$, and $-\frac{1}{3}f(x)$ on the same axis for $f(x) = 2^x$.

ANSWER:

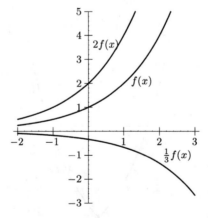

Figure 5.7

9. Fill in all the blanks in Table 5.3 for which you have sufficient information.

Table 5.3

x	-3	-2	-1	0	1	2	3
f(x)	4	2	-2	-3	0	4	3
f(-x)							
-f(x)							
f(x+3)							
f(x+3)							

ANSWER:

Table 5.4

x	-3	-2	-1	0	1	2	3
f(x)	4	2	-2	-3	0	4	3
f(-x)	3	4	0	-3	-2	2	4
-f(x)	-4	-2	2	3	0	-4	-3
f(x+3)	-3	0	4	3			
f(x)+3	7	5	1	0	3	7	6

10. Using Figure 5.8, graph the following functions.

   (a) $g(x) = 2f(x)$ **(b)** $h(x) = -f(x+1)$

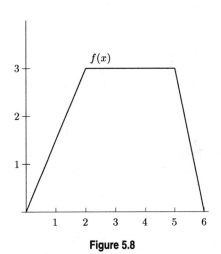

Figure 5.8

ANSWER:

(a)

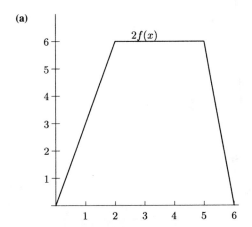

Figure 5.9

(b)

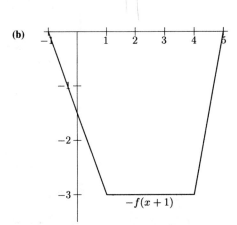

Figure 5.10

70

11. Without a calculator, graph the transformation $y = f(x + 1)$ of $f(x) = |x|$. Label at least three points.
ANSWER:

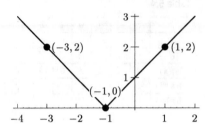

Figure 5.11

12. Using Table 5.5, Make a table of values for $f(2x)$ for an appropriate domain.

Table 5.5

x	-3	-2	-1	0	1	2	3
f(x)	3	2	-1	-2	0	1	4

ANSWER:

Table 5.6

x	-1.5	-1	-0.5	0	0.5	1	1.5
f(x)	3	2	-1	-2	0	1	4

13. For $f(x) = 2^x + x^2$, graph and label $f(x)$, $f\left(-\dfrac{1}{2}x\right)$, and $f(2x)$ on the same axes between $x = -5$ and $x = 5$.
ANSWER:

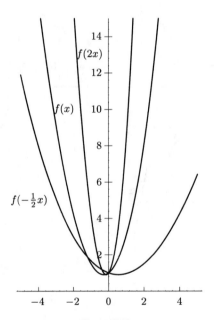

Figure 5.12

14. Using Figure 5.13, match each function to a graph (if any) that represents it:

(i) $y = f(-2x)$　　　　(ii) $y = f\left(\dfrac{1}{2}x\right)$　　　　(iii) $y = 2f(-x)$

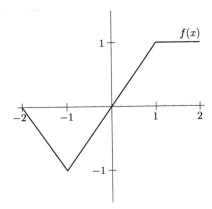

Figure 5.13

(a)

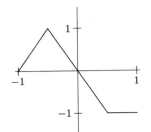

Figure 5.14

(b)

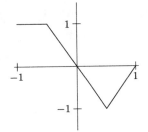

Figure 5.15

(c)

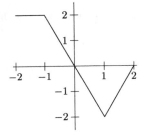

Figure 5.16

(d)

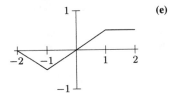

Figure 5.17

(e)

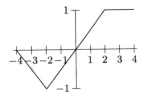

Figure 5.18

(f)

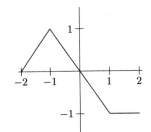

Figure 5.19

ANSWER:
(i) is (b), (ii) is (e), (iii) is (c).

15. Figure 5.20 gives a graph of $y = f(x)$, a quadratic function with vertex $(3, 2)$. Find a formula for $f(x)$.

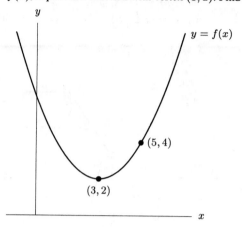

Figure 5.20

ANSWER:

Since a quadratic function can be placed in the form

$$y = a \cdot (x - h)^2 + k,$$

where (h, k) is the vertex, we have $f(x) = a \cdot (x - 3)^2 + 2$. Since $f(5) = 4$, this gives

$$a \cdot (5 - 3)^2 + 2 = 4$$
$$a \cdot (2)^2 = 2$$
$$4a = 2$$
$$a = \frac{1}{2}.$$

Thus,

$$f(x) = \frac{1}{2}(x - 3)^2 + 2.$$

16. Find a possible formula for $f(x)$:

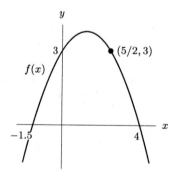

Figure 5.21

ANSWER:

$f(x) = k(x + 1.5)(x - 4)$, substituting $(\frac{5}{2}, 3)$ gives:

$$3 = k\left(\frac{5}{2} + \frac{3}{2}\right)\left(\frac{5}{2} - \frac{8}{2}\right)$$
$$3 = k\left(\frac{8}{2}\right)\left(-\frac{3}{2}\right)$$
$$3 = k(-6)$$
$$k = -\frac{1}{2}$$

so $f(x) = -\frac{1}{2}(x + 1.5)(x - 4)$.

17. Put $f(x) = x^2 + 6x - 2$ in vertex form by completing the square and graph without a calculator.

ANSWER:

To complete the square, first find the square of half of the coefficient of the x–term: $(6/2)^2 = 9$. Add and subtract this number:

$$f(x) = x^2 + 6x + 9 - 9 - 2$$
$$f(x) = (x + 3)^2 - 11.$$

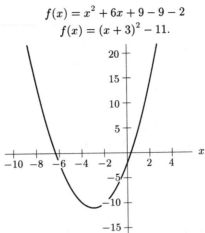

Figure 5.22

18. Show that the function $y = x^2 - 4x + 5$ has no real zeros.

ANSWER:

First, complete the square of the function by adding and subtracting the square of half of the coefficient of the x–term:

$$y = x^2 - 4x + 4 - 4 + 5$$
$$y = (x - 2)^2 + 1.$$

Now, try setting $y = 0$:

$$y = (x - 2)^2 + 1 = 0$$
$$(x - 2)^2 = -1$$

This is impossible, since the square of a number cannot be negative. Thus, y has no real zeros.

19. Suppose that a quadratic function has its vertex at $(3, 0)$ and has y-intercept -4. Find a formula for the function.

ANSWER:

The vertex is at $(h, k) = (3, 0)$. Substituting in this vertex in the standard equation for quadratic functions, we get

$$y = a(x - h)^2 + k = a(x - 3)^2.$$

When $x = 0$, $y = -4$, so $-4 = a(-3)^2 = 9a$, so $a = -\frac{4}{9}$. Thus,

$$y = -\frac{4}{9}(x - 3)^2.$$

Problems

20. Given the graph of $y = f(x)$ below, find a possible formula for $f(x)$.

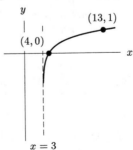

ANSWER:

$f(x) = \log(x - 3)$

21. One of these graphs shows a transformation of an exponential function, and one is a transformation of a log function. The remaining graph is not a transformation of either an exponential or log function. State which graph is which, and give reasons for your answers.

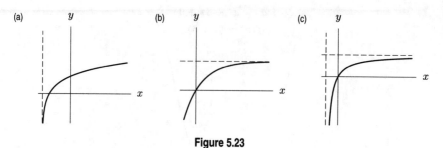

Figure 5.23

ANSWER:

(a) log, because it has one vertical asymptote

(b) exponential, because it has one horizontal asymptote

(c) neither, because it has both a horizontal and a vertical asymptote.

22. Let $g(x) = \sqrt{4 - x^2}$.

(a) Write a function that will translate the graph of g three units to the right and one unit up.

(b) Give the domain and range of the new function.

ANSWER:

(a) Let $h(x)$ denote the new function. Then to translate $g(x)$ 3 units to the right, we want

$$h(x) = g(x - 3)$$

for the set of all real x over which both h and g are defined. We thus replace x by $(x - 3)$ in the expression for $g(x)$:

$$h(x) = \sqrt{4 - (x - 3)^2}.$$

To translate the graph of g one unit up, we just add 1 to the expresion we have:

$$h(x) = 1 + \sqrt{4 - (x - 3)^2}.$$

(b) $h(x)$ is defined on the set of real numbers for all real values of x which make

$$4 - (x - 3)^2 \geq 0$$

so

$$(x - 3)^2 \leq 4$$

from which we get

$$-2 \leq x - 3 \leq 2$$
$$1 \leq x \leq 5.$$

This is the domain of $h(x)$. For these x values, $h(1) = 1$ is the minimum function value and $h(3) = 3$ is the maximum function value. Thus, for the given domain, the range of h is the set of all real numbers greater then or equal to 1 but less than or equal to 3.

23. A cup of coffee is initially at $170°F$, which is $100°F$ above room temperature. The difference between the coffee's temperature and room temperature decreases at the hourly rate of 80%. Find a formula for $T(t)$, the coffee's temperature after t hours have elapsed. [Hint: T will be a vertically-shifted exponential function.]

ANSWER:

The difference between the coffee's temperature and room temperature decays by 80% per hour. This means difference $= 100(0.2)^t$. So, $T(t) = 70 + 100(0.2)^t$

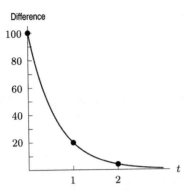

Figure 5.24: The difference in temperatures is exponential.

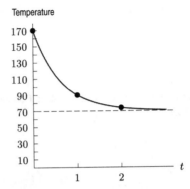

Figure 5.25: This graph is a vertically shifted exponential function.

24. Below is the graph of a function $f(x)$. Sketch a graph of $y = -f(x + 4) + 2$. Label at least three points on the graph.

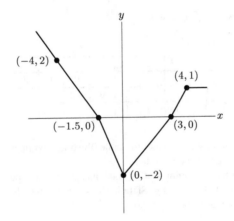

ANSWER:

We know that the graph of $y = -f(x + 4) + 2$ will look like the graph of $f(x)$ after having been left shifted by 4, reflected over the x-axis and then vertically shifted by 2. Thus the graph will look like

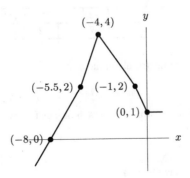

25. Figure 5.26 is the graph of a function $y = f(x)$. Make a careful, completely labeled sketch of the graph of $y = f(-x) + 2$.

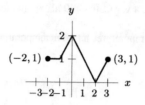

Figure 5.26

ANSWER:

To get $y = f(-x)$, we need to reflect f about the y axis. Then, by raising the resulting graph by 2, we get the graph in Figure 5.27.

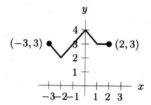

Figure 5.27

26. Table 5.7 gives values for the functions f, g, and h.

Table 5.7

x	-3	-2	-1	0	1	2	3
$f(x)$	4	3	2	1	0	1	2
$g(x)$	2	5	3	0	-3	-5	-2
$h(x)$	1	-7	4	-2			

(a) For each of the functions f and g, state whether the table illustrates symmetry about the x-axis, symmetry about the y-axis, symmetry about the origin, or none of these.

(b) Fill in the table for the function h, assuming that its graph has one of the types of symmetry (x-axis, y-axis, or origin) that is *not* illustrated by the table for f and g. State in the space below what type of symmetry is illustrated by the table for h after you have filled in the table

ANSWER:

(a) The function f has none of the discussed symmetries. The function g satisfies the property that $g(-x) = -g(x)$, and is symmetric about the origin.

(b)

Table 5.8

x	-3	-2	-1	0	1	2	3
$h(x)$	1	-7	4	-2	4	-7	1

We take the function h to be symmetric about the y-axis.

27. Let $g(x) = \dfrac{1}{x^2}$.

(a) Give the formula for a function that transforms the graph of g as follows:
   Shifts three units to the right,
   reflects about the x-axis,
   then shifts up two units.

(b) Give the domain and range of the new function.

(c) Is the new function even, odd, or neither? Why?

ANSWER:

(a) Let $y = \dfrac{1}{x^2}$. To shift the graph three units to the right, for any given x value, we want the value of y to be the value of $g(x-3)$. So we subtract 3 from x in the expression for $g(x)$:

$$y = \frac{1}{(x-3)^2}.$$

For reflection about the x-axis, since all values of y are positive, we just need to negate the y values:

$$y = -\frac{1}{(x-3)^2}.$$

To shift y us by two units, we add 2 to the expression for y:

$$y = 2 - \frac{1}{(x-3)^2}.$$

Let $f(x)$ denote the final expression.

(b) $f(x)$ is not defined when $x = 3$ so the domain is all real numbers x, $x \neq 3$. The range of $f(x)$ is all real numbers except 2.

(c) It is neither: it is not odd since $f(-x)$ is not equal to $-f(x)$ for all x, it is not even since $f(-x)$ is not equal to $f(x)$ for all x.

28. Figure 5.28 shows part of the graph of an even function. Complete the graph and the table of values for the function.

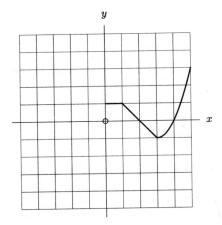

Figure 5.28

Table 5.9

x	y
1	1
2	0
3	-1
4	0
5	3

Table 5.10

x	y

ANSWER:

Let $f(x)$ denote the function. Since f is an even function,

$$f(-x) = f(x)$$

for all x in the domain of f. Therefore,

$$f(-1) = f(1) = 1$$
$$f(-2) = f(2) = 0$$
$$f(-3) = f(3) = -1$$
$$f(-4) = f(4) = 0$$
$$f(-5) = f(5) = 3$$

New table:

Table 5.11

x	y
-1	1
-2	0
-3	-1
-4	0
-5	3

The graph of f is symmetrical about the y-axis:

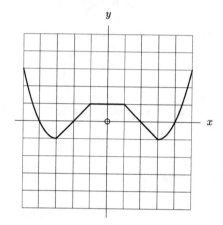

Figure 5.29

29. Figure 5.30 gives graphs of f, u, and v. The functions u and v are transformations of f. Find formulas in terms of f for u and v.

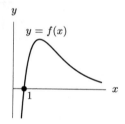 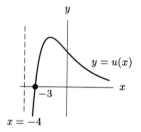 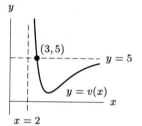

Figure 5.30

ANSWER:
$u = f(x + 4)$ and $v = 5 - f(x - 2)$

30. The graph of $f(x)$ is shown in Figure 5.31. Give a formula in terms of $f(x)$ for each of the graphs (a)–(d).

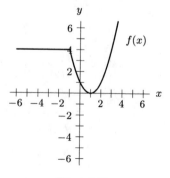

Figure 5.31

(a)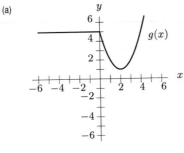

(b)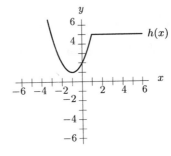

(c)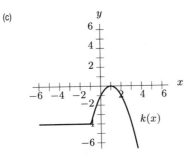

(d)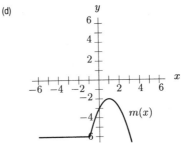

Figure 5.32

ANSWER:

(a) The graph of $f(x)$ is shifted up 1 (outside $+1$) and to the right 1 (inside -1):

$$g(x) = f(x-1) + 1.$$

(b) The graph of $f(x)$ is flipped horizontally over the y-axis and shifted up 1:

$$h(x) = f(-x) + 1.$$

(c) The graph of $f(x)$ is flipped vertically over the x-axis:

$$k(x) = -f(x).$$

(d) The graph of $f(x)$ is flipped vertically over the x-axis and then shifted down 2:

$$m(x) = -f(x) - 2.$$

31. The function $f(x)$ has odd symmetry and the function $g(x)$ has even symmetry. State whether the following functions are even, odd or neither.

(a) $h(x) = \dfrac{f(x)}{g(x)}$

(b) $k(x) = f(x+1)$

(c) $p(x) = g(x) + 1$

(d) $q(x) = -2f(x)$

ANSWER:

To check whether a function has even or odd symmetry, substitute $-x$, and examine the results. Since $f(x)$ is odd, $f(-x) = -f(x)$. Since $g(x)$ is even, $g(-x) = g(x)$. We can use these results to check the given functions:

(a) Odd symmetry. Substitute $-x$:

$$h(-x) = \frac{f(-x)}{g(-x)}$$
$$= \frac{-f(x)}{g(x)}$$
$$= -h(x).$$

Since $h(-x) = -h(x)$, h has odd symmetry.

(b) Neither. Substitute $-x$: $k(-x) = f(-x + 1)$
Since $k(-x) \neq -k(x)$, k does not have odd symmetry.
Since $k(-x) \neq k(x)$, k does not have even symmetry.

(c) Even symmetry. Substitute $-x$:

$$p(-x) = g(-x) + 1$$
$$= g(x) + 1$$
$$= p(x).$$

Since $p(-x) = p(x)$, p has even symmetry.

(d) Odd symmetry. Substitute $-x$:

$$q(-x) = -2f(-x)$$
$$= -2[-f(x)]$$
$$= 2f(x)$$
$$= -[-2f(x)]$$
$$= -q(x).$$

Since $q(-x) = -q(x)$, q has odd symmetry.

32. Let $f(x)$ be defined by the graph in Figure 5.33.

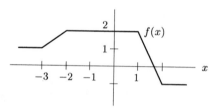

Figure 5.33

Find formulas in terms of $f(x)$ for the following transformations.

(a)

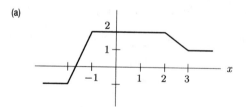

Figure 5.34

(b)

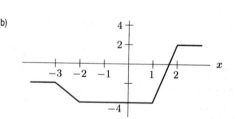

Figure 5.35

ANSWER:

(a) Figure 5.34 looks like $f(x)$ flipped across the y-axis. This means that the function value which once corresponded to a particular x now goes with its negative. Thus, $y = f(-x)$ is the new formula.

(b) Figure 5.35 is the same shape as $f(x)$ flipped over the x-axis. However, the y-values show that the function has also been vertically stretched by a factor of 2. Thus, the new formula is $-2f(x)$.

33. Figure 5.36 shows the graph of $f(x)$ and the graphs of several multiples of $f(x)$.

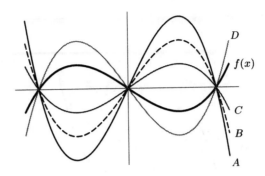

Figure 5.36

Suppose A is the graph of $y = a \cdot f(x)$,
B is the graph of $y = b \cdot f(x)$,
C is the graph of $y = c \cdot f(x)$,
D is the graph of $y = d \cdot f(x)$.

Circle the statements below which are true.

$$a < b$$

$$a < 0$$

$$b < c$$

$$c < d$$

$$c < 0$$

$$d < 0$$

$$d < 1$$

ANSWER:

Since graph D is a vertical stretch of the graph of $f(x)$ and is not reflected over the x-axis, $d > 1$. The graphs A, B, and C all represent vertical stretches of the graph of $f(x)$ which are reflected over the x-axis. Therefore, a, b, and c are negative. Since graph A is the "most stretched", a is the most negative. We can then see that $a < b < c < 0$.

$\boxed{a < b}$
$\boxed{a < 0}$
$\boxed{b < c}$
$\boxed{c < d}$
$\boxed{c < 0}$
$d < 0$
$d < 1$

34. Figure 5.37 shows the graph of $y = f(x)$. On the axes below, sketch a graph of $y = 2f(x)$. Label intercepts and at least one other point.

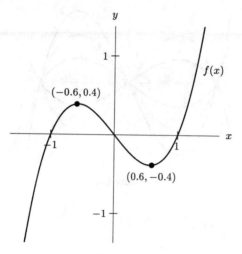

Figure 5.37

ANSWER:

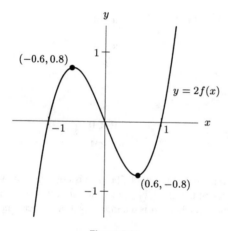

Figure 5.38

35. Figure 5.39 contains the graph of the function $f(x)$.

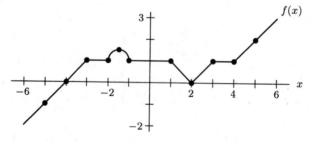

Figure 5.39

Figures 5.40 and 5.41 show graphs of functions which are transformations of the graph of $f(x)$. Find a formula for each of the functions in terms of $f(x)$.

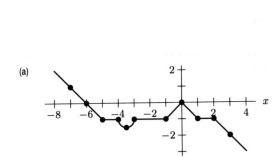

(a)

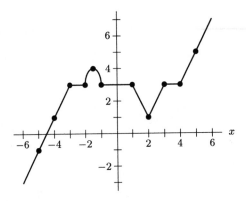

(b)

Figure 5.40

Figure 5.41

ANSWER:

(a) Figure 5.40 is Figure 5.39 translated left 2 units and reflected across the x-axis. Therefore, Figure 5.40 has the formula:
$$-f(x+2).$$

(b) Figure 5.41 was produced from Figure 5.39 by stretching vertically by a factor of 2 and translating up by one unit. So Figure 5.41 has the formula:
$$2f(x)+1.$$

36. The graph of $f(x)$ is in Figure 5.42.

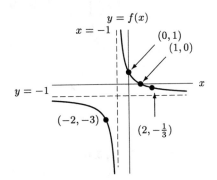

Figure 5.42

(a) Graph $y = 2 - f(x+3)$

(b) Graph $y = 3f(-2x)$

ANSWER:

(a) The graph of $y = 2 - f(x+3)$ will be the graph of f

- shifted left 3 units
- flipped across the x-axis
- shifted up 2 units.

The graph of $y = 2 - f(x+3)$ is shown in Figure 5.43.

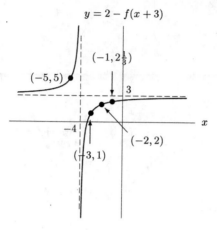

Figure 5.43

(b) The graph of $y = 3f(-2x)$ will flip the graph of f about the y-axis, compress the graph of f by a factor of 2 along the x-axis, and stretch the function by a factor of 3 along the y-axis. See Figure 5.44

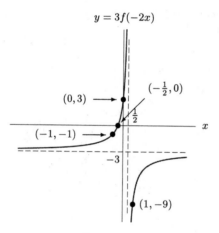

Figure 5.44

37. The graph of $y = r(x)$ is given in Figure 5.45.

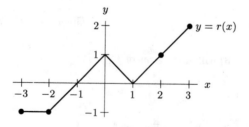

Figure 5.45: A graph of $y = r(x)$

(a) Sketch $y = 2 - r\left(\dfrac{x}{2}\right)$. Label both axes.

(b) Sketch $y = 2r(x + 1) - 2$. Label both axes.

ANSWER:

(a) Writing this formula as $y = -r\left(\dfrac{x}{2}\right) + 2$, we see that it describes the graph of r stretched horizontally by a factor of 2, then flipped across the x-axis, and then shifted up by 2 units. The resulting graph is given in Figure 5.46.

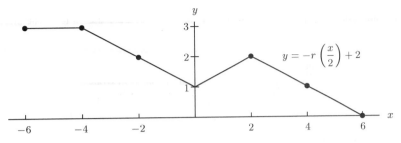

Figure 5.46: $y = -r\left(\frac{x}{2}\right) + 2$

(b) This formula describes the graph of r shifted left by 1 unit, then stretched vertically by a factor of 2, then shifted down by 2 units. The resulting graph is given in Figure 5.47.

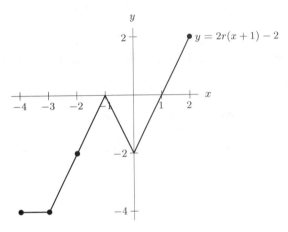

Figure 5.47: $y = 2r(x+1) - 2$

38. Educational psychologists suggest that a child's intellectual capacity, I can be represented as a function of time, t, in months, as follows

$$I = f(t) = a(1 - e^{-bt}),$$

where $a > 0, b > 0$ are constants.

(a) Does the graph of this function have a vertical asymptote? If so, what is it?

(b) Does the graph of this function have a horizontal asymptote? If so, what is it?

(c) What is the effect on the graph of I of increasing the value of a?

(d) What is the effect on the graph of I of increasing the value of b?

ANSWER:

(a) No.

(b) Yes; it is $I = a$.

(c) Increasing a moves the asymptote up.

(d) Increasing b makes the graph get close to the asymptote more quickly.

39. Table 5.12 gives values of $c(x)$, the cost per foot of drilling a well at a depth of x feet. The prices given are for wells drilled on a certain ranch in Texas.

Table 5.12

x	0	20	40	60	80	100	120	140
$c(x)$	75	75	75	120	125	135	80	80

(a) Estimate the cost of drilling a well that is 40 feet deep.

(b) The ranch sits above a dense layer of bedrock. At what depth do you estimate the bedrock to begin? (Give a range.)

(c) Let $c_1(x)$ and $c_2(x)$ give the drilling costs per foot on two nearby ranches. The bedrock layer that lies beneath the first ranch extends beneath these ranches, too. (No reasons need be given.)

    (i) If $c_1(x) = c(x + 20)$, does the bedrock layer beneath this ranch lie closer to or further from the surface than it does at the first ranch?

    (ii) If $c_2(x) = c(1.25x)$, is the bedrock layer beneath this ranch thicker or less thick than it is beneath the first ranch?

ANSWER:

(a) The cost per foot (for the first 40 feet) seems to be a constant \$75/ft. Thus at a depth of 40 feet, the cost is $40(75) =$ \$3000.

(b) The cost jumps dramatically between $x = 40$ and $x = 60$. This implies that the bedrock layer begins somewhere in this range.

(c)   **(i)** Closer to the surface because the cost data shifts to the smaller (more shallow) x-values. That is, a graph of cost versus depth would be shifted to the left.

    (ii) The graph of c_2 would resemble a horizontally compressed version of the graph of c_1. Thus the jump in the c-values between $x = 60$ and $x = 100$ (on the original graph) would be less spread out and so the bedrock layer is less thick here.

40. Let $y = p(x)$ be defined by the following graph, and let $y_0 = p(x_0)$ and $y_1 = p(x_1)$.

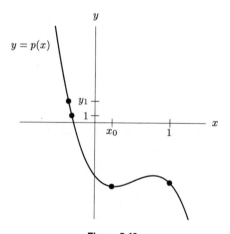

Figure 5.48

Rank each of the following lists of expressions in order from least to greatest.

(a) $0, \quad 1, \quad x_0, \quad x_1, \quad y_0, \quad y_1$

(b) $p(x_0), \quad p(x_0 + 1), \quad p(x_0) + 1, \quad p(x_0 - 1)$

(c) $p(x_0), \quad p(2x_0), \quad 2p(x_0), \quad p(-x_0), \quad -p(x_0)$

ANSWER:

(a) $y_0 < x_1 < 0 < x_0 < 1 < y_1$

(b) $p(x_0 + 1) < p(x_0) < p(x_0) + 1 < p(x_0 - 1)$

(c) $2p(x_0) < p(x_0) < p(2x_0) < p(-x_0) < -p(x_0)$

41. Suppose that $h(x) = x^2 + 3$.

   (a) Given no information other than this formula, find the domain and range of the function, stating clearly which is which.

   (b) Now suppose that you are told that the quantity $h(x)$ represents the number of hours it takes a farmer to fertilize a square field with side length x (in thousands of feet). With this additional information, find the domain and the range of the function, stating clearly which is which.

   ANSWER:

   (a) With no other information, we have to assume that x has no restrictions, so the domain is $-\infty < x < \infty$. Since $x^2 \geq 0$, we know that $h(x) = x^2 + 3 \geq 3$. Thus, the range is $3 \leq h(x) < \infty$.

   (b) In the context of the farmer who is fertilizing a square field, it does not make sense for the farmer to be able to fertilize a field of negative side length. Thus, the domain is $x \geq 0$. Given that domain, the range is unchanged. Thus, the range is $3 \leq h(x) < \infty$.

42. A relief package is dropped from an airplane and falls to the ground along a parabolic path, starting at the vertex of the parabola. The package was released when the plane was directly above a marker at a height of 5 kilometers, and the package hits the ground 4.43 kilometers from the marker. If h is the height of the package when it is a horizontal distance d from the marker, find an equation for h in terms of d.

   ANSWER:

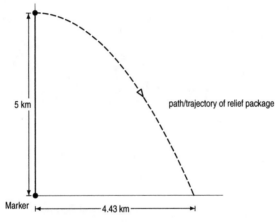

Figure 5.49

The path along which the package falls is a parabola and can thus be described by a quadratic function. h can therefore be expressed as a quadratic function of d:

$$h = ad^2 + bd + c$$

where a, b, and c are constants. At the instant the package is released, $d = 0$. So,

$$5 = a(0)^2 + b(0) + c$$
$$c = 5$$

When the package gets to the ground, $d = 4.43$ and $h = 0$, so

$$0 = a(4.43)^2 + b(4.43) + 5$$
$$19.6a = -5 - 4.43b.$$

Since the vertex of the parabola occurs at $d = 0, b = 0$. So

$$a = \frac{-5}{4.43^2} \approx -0.2548.$$

Thus, $h = -0.2548d^2 + 5$.

43. A projectile's height $h(t)$ in feet above ground is a quadratic function of time t in seconds since launched. Three values of the function are given in the table below.

t	$h(t)$
0	64
2	96
3	64

(a) What is the practical interpretation of $h(0)$?

(b) Find a formula for $h(t)$.

(c) Find when the projectile strikes the ground.

(d) Find the vertex of the parabola.

(e) What is the physical interpretation of the vertex in the context of this problem?

ANSWER:

(a) $h(0)$ is the height of the projectile above ground at the instant it is fired.

(b) Since $h(t)$ is a quadratic function, let

$$h(t) = at^2 + bt + c$$

where a, b, and c are constants. From the table,

$$h(0) = 64 = a(0)^2 + b(0) + c$$
$$c = 64$$
$$h(2) = 96 = a(2)^2 + b(2) + 64$$
$$32 = 4a + 2b$$
$$16 = 2a + b$$
$$h(3) = 64 = a(3)^2 + b(3) + 64$$
$$0 = 9a + 3b$$
$$b = -3a.$$

Substituting $b = -3a$ in $16 = 2a + b$ gives $2a - 3a = 16$, so $a = -16$ and then $b = -3(-16) = 48$. Thus, we have

$$h(t) = -16t^2 + 48t + 64$$

(c) When the projectile strikes the ground, its height above the ground is zero, so

$$h(t) = 0 = -16t^2 + 48t + 64$$
$$-t^2 + 3t + 4 = 0$$
$$-t^2 - t + 4t + 4 = 0$$
$$-t(t + 1) + 4(t + 1) = 0$$
$$(t + 1)(-t + 4) = 0$$
$$t = -1 \text{ or } t = 4.$$

We want the time it strikes the ground after the projectile is launched, so t should be greater than zero. Thus, $t = 4$ seconds.

(d) The vertex of the parabola is at $\left(\dfrac{-b}{2a}, f\left(\dfrac{-b}{2a}\right)\right)$.

$$f\left(\frac{-b}{2a}\right) = f\left(\frac{-48}{2(-16)}\right)$$
$$= f\left(\frac{3}{2}\right)$$
$$= -16\left(\frac{3}{2}\right)^2 + 48\left(\frac{3}{2}\right) + 64$$
$$= -4(9) + 24(3) + 64$$
$$= 100 \text{ feet.}$$

Alternatively, since h is quadratic and we have two values of t for which the corresponding h values are equal, $t = 0$ and $t = 3$, the vertex will occur midway between these two points so t will be $\dfrac{0+3}{2} = 1.5$ and we could go ahead and find $f(1.5)$ to get the vertex. So the vertex occurs at the point $(\frac{3}{2}, 100)$.

(e) The vertex is the highest point above ground to which the projectile rises.

44. Using a calculator or computer, sketch the graphs of $f(x) = x^2$ and $g(x) = x^2 + 10x + 18$ as they appear on the window $-10 \le x \le 10, -10 \le y \le 10$.

(a) Describe in words how the second graph compares to the first.

(b) Using transformations, find a formula for $g(x)$ in terms of $f(x)$.

(c) Prove your formula is correct by completing the square for $g(x)$.

ANSWER:

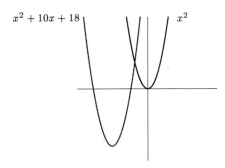

Figure 5.50

(a) The second graph is shifted left by 5 units and moved down 7 units.

$$g(x) = (x+5)^2 - 7$$

(b) $g(x)$ in terms of $f(x)$:

$$f(x) = x^2$$
$$g(x) = f(x+5) - 7$$

(c)

$$\begin{aligned} g(x) &= x^2 + 10x + 18 \\ &= (x^2 + 10x) + 18 \\ &= (x^2 + 10x + 25) + 18 - 25 \\ &= (x+5)^2 - 7 \end{aligned}$$

45. (a) Find a formula for the parabola $f(x)$ in Figure 5.51.

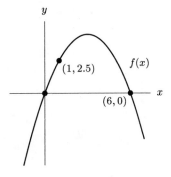

Figure 5.51

(b) Put your formula in vertex form: $f(x) = a(x-h)^2 + k$. Show all work.

ANSWER:

(a) Since the x-intercepts are at $x = 0$ and $x = 6$, the quadratic function $f(x)$ must be of the form $f(x) = ax(x - 6)$. Substituting $x = 1, y = 2.5$, we have

$$2.5 = a(1)(1 - 6)$$
$$a = \frac{2.5}{-5} = -0.5$$

Therefore, $f(x) = -0.5x(x - 6)$.

(b) Due to the symmetry of a quadratic function, the x-coordinate of the vertex must be $x = 3$. Since $f(3) = -0.5(3)(-3) = 4.5$, the vertex $(h, k) = (3, 4.5)$. Thus,

$$f(x) = -0.5(x - 3)^2 + 4.5.$$

Chapter 6 Exam Questions

Exercises

1. Does the following function appear to be periodic with period less than 4?

Table 6.1

x	0	1	2	3	4	5	6	7
f(x)	0	1	0	2	0	3	0	4

ANSWER:
No

2. Does the following function appear to be periodic with period less than 4π?

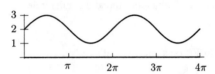

Figure 6.1

ANSWER:
Yes

3. Estimate the period of the following periodic function.

Table 6.2

r	π	2π	3π	4π	5π	6π	7π	8π	9π
h(r)	-1	0	1	0	-1	0	1	0	-1

ANSWER:
4π

4. Estimate the period of the following periodic function.

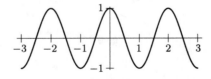

Figure 6.2

ANSWER:
2

5. What angle (in degrees) corresponds to -1.5 rotations around the unit circle?
ANSWER:
$-1.5(360°) = -540°$

6. Find the coordinates of the point at the given angle on the unit circle.

    (a) 60°   **(b)** −210°

    ANSWER:

    (a) The coordinates are $(\cos 60°, \sin 60°)$, which is approximately $(0.5, 0.866)$.

    (b) The coordinates are $(\cos(-210°), \sin(-210°))$, which is approximately $(-0.866, 0.5)$.

7. Find the coordinates of the point at the given angle on a circle of radius 2.5 centered at the origin.

    (a) 400°   **(b)** −75°

    ANSWER:

    (a) The coordinates are $(2.5\cos(400°), 2.5\sin(400°))$, which is approximately $(1.915, 1.607)$.

    (b) The coordinates are $(2.5\cos(-75°), 2.5\sin(-75°))$, which is approximately $(0.647, -2.415)$.

8. Convert the angle 135° to radians

    ANSWER:

$$135° \cdot \frac{\pi \text{ radians}}{180°} = \frac{135\pi}{180} = \frac{3\pi}{4}$$

9. Convert the angle $-\dfrac{5\pi}{4}$ (radians) to degrees.

    ANSWER:

$$-\frac{5\pi}{4} \cdot \frac{180°}{\pi \text{ radians}} = -225°$$

10. What angle in radians corresponds to -1.25 rotations around the unit circle?

    ANSWER:

$$-1.25(2\pi) = -2.5\pi.$$

11. Find the arc length corresponding to 30° on a circle of radius 3?

    ANSWER:

$$30° = \frac{\pi}{6}, \text{ so the arc length is } 3\left(\frac{\pi}{6}\right) = \frac{\pi}{2}.$$

12. Find the midline and amplitude of the periodic function $f(x) = 3\cos(4x) - 1$.

    ANSWER:

    The midline is $y = -1$; the amplitude is 3.

13. Find the midline and amplitude of the periodic function

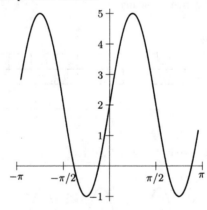

Figure 6.3

    ANSWER:

    The midline is $y = 2$; the amplitude is 3.

14. Find the exact value of $\cos\left(\dfrac{2\pi}{3}\right)$ without a calculator.

    ANSWER:

$$\cos\left(\frac{2\pi}{3}\right) = -\cos\left(\frac{\pi}{6}\right) = -\frac{1}{2}$$

15. State the period, amplitude, and midline of $y = 3\sin(2t - 1) + 4$.

    ANSWER:

    Period is $\dfrac{2\pi}{2} = \pi$. Amplitude is 3. Midline is $y = 4$.

16. What are the horizontal and phase shifts of $y = 4\cos(2t + 6) - 7$.

   ANSWER:

   First, rewrite the equations as $y = 4\cos(2(t+3)) - 7$. The horizontal shift is -3. The phase shift is -6.

17. Find a formula for the trigonometric function.

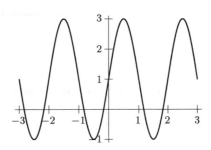

Figure 6.4

   ANSWER:

   The function resembles the graph of $f(t) = \sin t$, with the following transformations: The period of the function is 2, the midline is $y = 1$, the amplitude is 2, and there is no horizontal shift. Thus, the function is $f(t) = 2\sin(\pi t) + 1$.

18. Without a calculator, find the exact value of $\tan 45°$.

   ANSWER:

   $$\tan 45° = \frac{\sin 45°}{\cos 45°} = \frac{\frac{\sqrt{2}}{2}}{\frac{\sqrt{2}}{2}} = 1$$

19. Without a calculator, find the exact value of $\sec 120°$.

   ANSWER:

   $$\sec 120° = \frac{1}{\cos 120°} = \frac{1}{-\frac{1}{2}} = -2$$

20. Without a calculator, find the exact value of $\tan\left(-\frac{\pi}{6}\right)$.

   ANSWER:

   $$\tan\left(-\frac{\pi}{6}\right) = \frac{\sin\left(-\frac{\pi}{6}\right)}{\cos\left(-\frac{\pi}{6}\right)} = \frac{-\frac{1}{2}}{\frac{\sqrt{3}}{2}} = -\frac{\sqrt{3}}{3}.$$

21. Without a calculator, find the exact value of $\csc\left(\frac{3\pi}{4}\right)$.

   ANSWER:

   $$\csc\left(\frac{3\pi}{4}\right) = \frac{1}{\sin\left(\frac{3\pi}{4}\right)} = \frac{1}{\frac{\sqrt{2}}{2}} = \sqrt{2}.$$

22. **(a)** Use a graph of $y = \sin t$ to estimate two solutions to the equation $\sin t = 0.25$ for $0 \le t \le \pi$.

   (b) Solve the same equation using inverse sine.

   ANSWER:

   (a) Looking at the graph, we see solutions at $t = 0.25$ and $t = 2.89$.

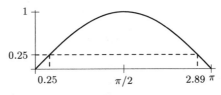

Figure 6.5

   (b) $\arcsin(0.25) = 0.253$, and the other solution would be at $t = \pi - 0.253 = 2.889$.

23. Find a solution with θ in radians for $\tan\theta = 50$.

   ANSWER:

   1.551

24. Without a calculator, find exact values for:

(a) $\cos 300°$

(b) $\sin\left(-\dfrac{3\pi}{4}\right)$

ANSWER:

(a) $\cos 300° = \dfrac{1}{2}$

(b) $\sin\left(-\dfrac{3\pi}{4}\right) = -\dfrac{\sqrt{2}}{2}$

25. Give the reference angle for the following:

(a) $\frac{3\pi}{4}$ (b) $\frac{5\pi}{6}$ (c) $-\frac{\pi}{6}$ (d) $\frac{13\pi}{6}$ (e) $\frac{7\pi}{6}$

ANSWER:

(a) $\pi/4$ (b) $\pi/6$ (c) $\pi/6$ (d) $\pi/6$ (e) $\pi/6$

Problems

26. A ferris wheel is 30 meters in diameter, and must be boarded from a platform that is 2 meters above the ground. The wheel completes one full revolution every 7 minutes. At the initial time $t = 0$ you are in the twelve o'clock position.

(a) Draw a carefully labeled graph of your height $h(t)$ above ground level t minutes after the inital time. Be sure to include at least two full cycles (period lengths) in this graph.

(b) State the period, the amplitude, and the midline of the graph drawn in part (a), stating which is which.

ANSWER:

(a)

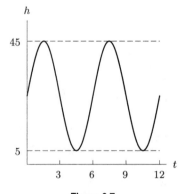

Figure 6.6: Graph of $y = h(t)$

(b) The period is 7, the amplitude is 15, and the midline is $y = 17$.

27. The graph shows your height $h = f(t)$ in meters t minutes after a ferris wheel ride begins.

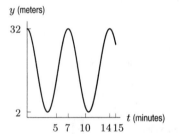

Figure 6.7

(a) How high is the hub of the ferris wheel?

(b) What is the radius of the ferris wheel?

(c) How many minutes are required for one complete revolution of the ferris wheel?

ANSWER:

(a) The hub is the point about which the wheel rotates, so you attain the highest height when directly above the hub, the lowest height directly below the hub, and midway between between when on the same horizontal level as the hub.

From the graph, we see an oscillation representing the motion of a point on the rim of the wheel, i.e. your motion. The oscillation occurs about the point at height 25m. This is the distance from the hub to the rim of the wheel plus the distance from the lowest point on the rim to the ground. Thus 25m. is the height of the hub above the ground.

(b) The radius of the wheel is the distance from the hub to any point on the rim of the wheel. Since we know the hub is 25m high and the lowest point on the rim of the wheel is 5m high, the radius is the difference

$$25m - 5m = 20m.$$

(c) One complete revolution is represented by one full cycle of the wave (one complete peak and one complete trough) in the figure. We can see from the figure that a cycle occurs over a period of 6 minutes so that is the time for a complete revolution.

28. The proposed London ferris wheel will be 500 feet in diamter and will make one revolution every 20 minutes. Let $y = h(t)$ be the height above ground in terms of t minutes of riding.

(a) What is the period, the amplitude, and the midline for $h(t)$?

(b) Describe in a sentence or two what the following expressions would mean in terms of the London ferris wheel (for example, bigger wheel or faster wheel etc.)

  (i) $h(t) + 5$

  (ii) $h(t + 5)$

  (iii) $h(5t)$

  (iv) $5h(t)$

  ANSWER:

(a) Since the wheel makes one full revolution in 20 minutes, the period of $h(t)$ must be 20. The amplitude is half the difference between the maximum and the minimum of $h(t)$. In our case, the minimum height is 0 and the maximum height is 500 ft so the amplitude is 250 ft. Likewise, the midline is the average of the maximum and minimum of $h(t)$ and also occurs at 250 ft

(b) (i) The loading platform is 5 feet off of the ground.

  (ii) This represents the height of a friend who boarded the ferris wheel 5 minutes before you.

  (iii) This represents the height on a wheel that runs 5 times faster.

  (iv) This represents the height on a wheel which is 5 times larger in diameter.

29. Suppose $f(a) = 3b$ and $f(2a) = 5b$. Find possible values of $f(0)$ and $f(3a)$ assuming that f is

(a) linear,

(b) exponential,

(c) periodic with a period of $2a$ (that is, the values of f repeat on intervals of length $2a$).

Hint: You don't need to find formulas to answer this question.

  ANSWER:

(a) If f is linear, then the difference (Δf) between successive terms in the sequence $f(0), f(a), f(2a), f(3a)$ will be constant. We see that

$$\Delta f = f(2a) - f(a) = 5b - 3b = 2b.$$

Thus,

$$f(0) = f(a) - \Delta f = 3b - 2b = b \qquad \text{and} \qquad f(3a) = f(2a) + \Delta f = 5b + 2b = 7b.$$

Notice that the resulting sequence of f-values goes up by equal steps:

$$b, 3b, 5b, 7b.$$

(b) If f is exponential, then the *ratio* of successive terms in the sequence $f(0), f(a), f(2a), f(3a)$ will be constant. We see that this ratio is given by

$$\frac{f(2a)}{f(a)} = \frac{5b}{3b} = \frac{5}{3}.$$

Thus, the ratio of $f(a)$ and $f(0)$ can be used to find $f(0)$. That is,

$$\frac{f(a)}{f(0)} = \frac{5}{3}$$

$$f(0) = \left(\frac{3}{5}\right) \cdot 3b = \frac{9}{5}b.$$

Similarly, we have

$$f(3a) = f(2a) \cdot \frac{5}{3} = 5b \cdot \frac{5}{3} = \frac{25}{3}b.$$

Notice that each number in the sequence of f-values is 5/3 times the previous number:

$$\frac{9}{5}b, 3b, 5b, \frac{25}{3}b.$$

(c) If the value of f repeats on intervals of $2a$, then $f(0) = f(2a) = f(4a) = \ldots$ and $f(a) = f(3a) = f(5a)\ldots$. This means that $f(0) = 5b$ and $f(3a) = 3b$. Notice that the resulting sequence of f-values alternates up and down periodically:

$$5b, 3b, 5b, 3b.$$

30. Find angle θ and the coordinates of point P on the unit circle in Figure 6.8.

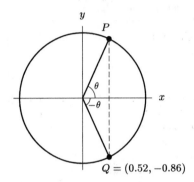

Figure 6.8

ANSWER:
By trial and error, we find $\cos 59° \approx 0.52$ and $\sin 59° \approx 0.86$. Thus, $\theta \approx 59°$.
The coordinates of $P = (\cos 59°, \sin 59°) = (0.52, 0.86)$.

31. Find the angles θ and ϕ and the coordinates of point R on the unit circle in Figure 6.9.

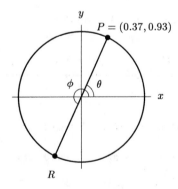

Figure 6.9

ANSWER:
By trial and error, we find $\cos 68° \approx 0.37$ and $\sin 68° \approx 0.93$. Thus, $\theta \approx 68°$, so $\phi = 180 + 68 = 248°$. The coordinates of $R = (\cos 248°, \sin 248°) = (-0.37, -0.93)$.

32. Find angle θ from the unit circle in Figure 6.10.

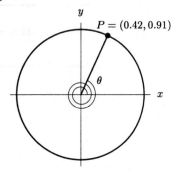

Figure 6.10

ANSWER:

By trial and error, we find $\cos 65° \approx 0.42$ and $\sin 65° \approx 0.91$. So θ is two complete revolutions plus $65°$, so $\theta = 2(360°) + 65° = 785°$.

ANSWER:

$785°$

33. (a) Find the coordinates of the points P and Q on the unit circle in Figure 6.11.
   (b) What is the vertical distance from P to Q?
   (c) What is the horizontal distance from Q to the y-axis?

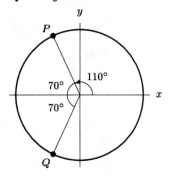

Figure 6.11

ANSWER:

(a) Coordinates of $P = (\cos 110°, \sin 110°) = (-0.342, 0.940)$.
   Therefore coordinates of Q are $(-0.342, -0.940)$. Alternatively coordinates of $Q = (\cos 250°, \sin 250°)$.
(b) $2\sin 110° = 1.88$.
(c) $|\cos 250°| = |\cos 110°| = 0.342$.

34. Figure 6.12 gives the unit circle and the function $f(x) = \frac{1}{x}$. Find the y-coordinate of the point P in terms of θ.

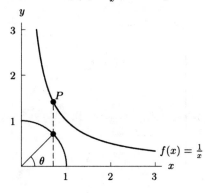

Figure 6.12

ANSWER:

If we know the x-coordinate of P we then know the y-coordinate of P. Since, if the x-coordinate of P is x we know that the y-coordinate of P must be $\frac{1}{x}$. On the other hand we know that the x-coordinate of P is simply the distance from the y-axis of the point on the arc determined by the angle θ. But that simply means that the x-coordinate of the point P is $\cos\theta$.

Thus the y-coordinate of P is

$$g(\theta) = \frac{1}{x\text{-coordinate of }P} = \frac{1}{\cos\theta}.$$

35. The Pentagon proposes to install a 5-panel revolving door with a panel width of 1 meter. A 5-star general wants to enter with a chart that is as wide as possible, what is the widest chart he can fit in the opening without bending? In other words, what is the distance across the opening?

ANSWER:

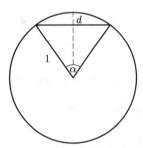

Figure 6.13

The angle α between door panels is $360°/5 = 72°$. Looking at the two right triangles we note that the lengths of their hypotenuses are both 1 meter. The distance, d, across the openings is the sum of the lengths of the two sides of the triangles opposite the angle α. Thus we must find the length of the opposite side when we know that the hypotenuse is of length 1 meter and the angle is $72°/2 = 36°$. The angle between the hypotenuse and the x-axis is $90° - 36° = 54°$. So we get

$$d = 2\cos(54°) \approx 1.176 \text{ meters.}$$

36. Figure 6.14 shows the path taken by the left front tire of a Honda Accord as the car changes direction sharply from due north to due east. The turning radius of the Accord is 18 feet, and the two front wheels are 58 inches apart. How far does the left front wheel travel during the turn from A to B? How far does the right front wheel travel during the same time?

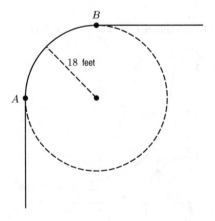

Figure 6.14

ANSWER:

We know that the left front tire travels one quarter of the circumference of a circle with radius 18 feet. That is, the left front tire travels

$$\frac{1}{4}(2\pi 18) = 9\pi \text{ ft.}$$

We know that the right front tire travels one quarter of the circumference of a circle with radius (18 feet − 58 inches). 58 inches is $4\frac{5}{6}$ feet, or $\frac{29}{6}$ feet, so the distance the right front tire travels is

$$\frac{1}{4}\left(2\pi\left(18 - \frac{29}{6}\right)\right) = \frac{1}{4}\left(2\pi\frac{79}{6}\right) = \left(\frac{79\pi}{12}\right) \text{ ft.}$$

37. **(a)** If $\theta = 37°$, what is the measure of θ in radians?
   (b) The minute hand on a watch is 0.8 inches long. How far (in inches) does the tip of the minute hand travel as the hand turns through $37°$?
   (c) How fast (in inches/minute) is the tip of the minute hand moving in (b)?
   ANSWER:

   (a) $37° \cdot \frac{\pi}{180°} \approx 0.65$ radians
   (b) Arc length= $0.8(0.65) = 0.52$
   (c) In one minute the hand moves $6°$ which is a length of $0.8 \cdot 6 \cdot \frac{\pi}{180} \approx 0.084$ inches/minute.

38. In the revolving door in Figure 6.15, (each panel is one meter) there are two important distances to measure between B and C.

   (a) What is the arc length between B and C?
   (b) What is the distance along a straight line between B and C?

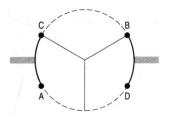

Figure 6.15

ANSWER:

(a) We know that the angle between the door panels at B and C sweeps out $120°$. Since the circumference of the circle is 2π meters, the length of the arc is

$$\frac{120°}{360°}2\pi = \frac{2\pi}{3} \approx 2.09 \text{ meters.}$$

(b)

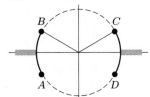

Figure 6.16

Setting the figure on coordinate axis, so that the circle is a unit circle, we get that the coordinate of point B are $(\cos 150°, \sin 150°)$ and the coordinates of point C are $(\cos 30°, \sin 30°)$. Thus the direct horizontal distance from point B to point C is

$$\begin{aligned}
\text{distance} &= \sqrt{(\cos 30° - \cos 150°)^2 + (\sin 30° - \sin 150°)^2} \\
&= \sqrt{(\cos 30° - (-\cos 30°))^2 + (\sin 30° - \sin 30°)^2} \\
&= \sqrt{(2\cos 30°)^2} \\
&= 2\cos 30° \\
&\approx 1.732 \text{ meters}
\end{aligned}$$

39.

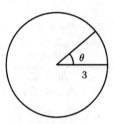

Figure 6.17

The sector of the circle in Figure 6.17 (the pizza-shaped piece) has an area which is the same fraction of the area of the whole circle as the angle θ is of a complete rotation. Find the area, A, of the sector as a function of θ. Assume θ is in radians.

ANSWER:

We know that the area of a circle of radius r is πr^2 and the angle of a complete rotation is 2π. So, in this case,

$$\frac{\text{Area of sector}}{\text{Area of circle}} = \frac{\text{Angle of sector}}{\text{Complete rotation}}$$

$$\frac{A}{\pi\,(3^2)} = \frac{\theta}{2\pi}$$

$$A = \frac{9}{2}\theta.$$

40. Let $f(x) = A\sin(x) + B$, with x in radians. Sketch graphs of $y = f(x)$, labeling both axes, assuming:

(a) $B > A > 0$ (b) $A > B > 0$

(c) $B > |A|$, and $B > 0 > A$ (d) $0 > A > B$

(e) $0 > B > A$ (f) $|B| > A$, and $A > 0 > B$

ANSWER:

(a)

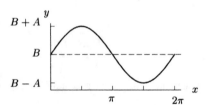

Figure 6.18

(b)

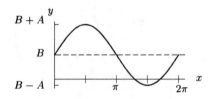

Figure 6.19

(c)

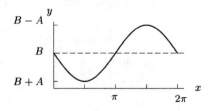

Figure 6.20

(d)

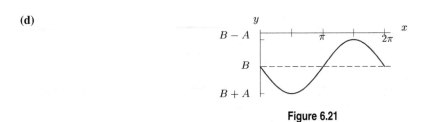

Figure 6.21

(e)

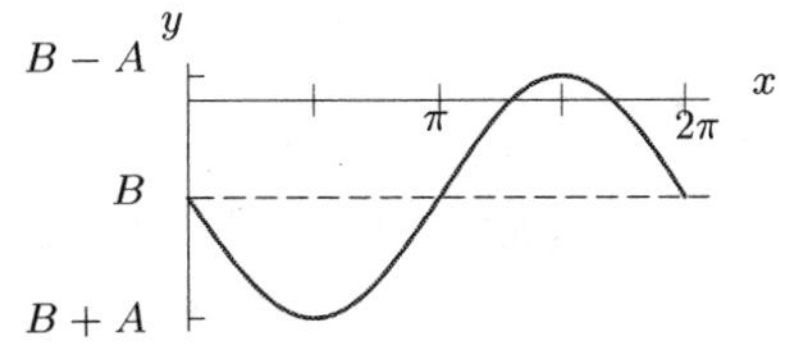

Figure 6.22

(f)

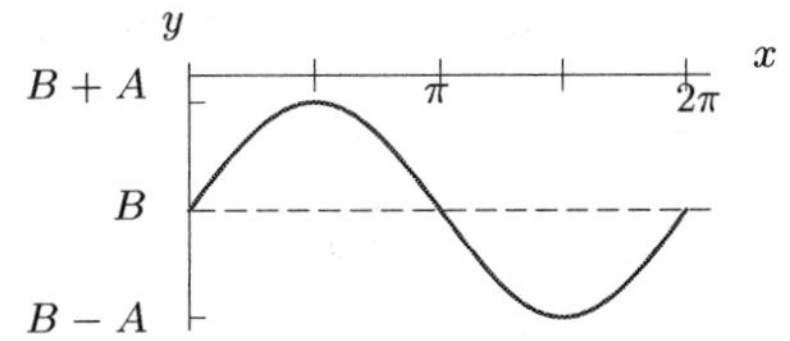

Figure 6.23

41. Suppose you are on a ferris wheel (that turns in a counter-clockwise direction) and that your height, in meters, above the ground at time, t, in minutes is given by

$$h(t) = 15 \sin(\frac{\pi}{2}t) + 15$$

(a) How high above the ground are you at time $t = 0$?
(b) What is your position on the wheel at $t = 0$? (That is, what o'clock?)
(c) What is the radius of the wheel?
(d) How long does one revolution take?

   ANSWER:

(a) 15 meters
(b) 3 o'clock or 9 o'clock position
(c) 15 meters
(d) Sketching a graph, we find values start to repeat after 4 minutes, so period is 4 minutes.

42. Which of the following equations could represent the graph in Figure 6.24? Circle all that apply.

(a) $y = A\cos(x + C) + D$, where $A > 0$, $C = 0$, and $D \leq 0$.
(b) $y = A\cos(x + C) + D$, where $A < 0$ and $C = D = 0$.
(c) $y = A\sin(x + C) + D$, where $A \leq D < 0$ and $C > 0$.
(d) None of the above.

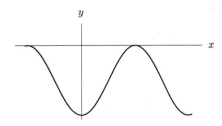

Figure 6.24

ANSWER:

Answer (a) does not apply since the cosine graph would have to be flipped or shifted horizontally to match the given graph. The condition that $A > 0$ prevents the graph from being flipped across the x-axis. The condition $C = 0$ prevents it from being shifted horizontally. Answer (b) does not apply since the cosine graph would have to be shifted downward, and $D = 0$ prevents that from happening. Answer (c) applies. In Figure 6.25 we see the graph $y = \sin x$. Since $A < 0$, Figure 6.26 shows the sine graph reflected across the x-axis. In Figure 6.27, we see the graph shifted down due to the fact that $D < 0$. Finally, because $C > 0$ we see the graph shifted to the left in Figure 6.28.

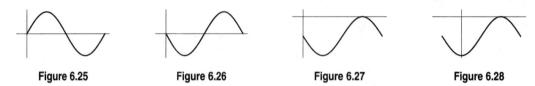

| **Figure 6.25** | **Figure 6.26** | **Figure 6.27** | **Figure 6.28** |

Answer (d) does not apply.

43.

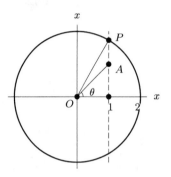

Figure 6.29

(a) Find the length of segment $\overline{OA}$ as a function of θ.

(b) Find the angle θ when A is at position P.

ANSWER:

(a) By the definition of cosine,

$$\cos\theta = \frac{1}{\overline{OA}}.$$

Thus,

$$\overline{OA} = \frac{1}{\cos\theta}.$$

(b) Again, by definition the angle θ such that A is at position P must satisfy

$$\cos\theta = \frac{1}{2}$$

But since in the Figure 6.29 the angle θ is in the first quadrant we know that

$$\theta = 60°.$$

44. Find a possible formula for the following trigonometric function.

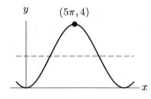

Figure 6.30

ANSWER:

The graph looks like that of an inverted cosine curve which has been stretched and vertically shifted. Thus the formula for the function would be

$$f(x) = -A\cos(Bx) + k$$

where A is the amplitude, $\frac{2\pi}{B}$ is the period and k is the midline.

We know that the the minimum of the function is $y = 0$ and the maximum is $y = 4$ thus

$$\text{Amplitude} = A = \frac{\text{max} - \text{min}}{2} = \frac{4 - 0}{2} = \frac{4}{2} = 2$$

and

$$\text{Midline} = k = \frac{\text{max} + \text{min}}{2} = \frac{4 + 0}{2} = \frac{4}{2} = 2.$$

We know that the function completes half its period at $x = 5\pi$ so the period of the function is 10π. Thus we get

$$\frac{2\pi}{B} = 10\pi$$

or in other words

$$B = \frac{1}{5}.$$

So we get

$$f(x) = -2\cos\left(\frac{1}{5}x\right) + 2.$$

45. City A is a modest tourist town, which means that its population undergoes a seasonal variation. In January, it dips down to 4,500 people, but by July, with the warm weather, its population climbs to around 5,500 people. By the following January, however, the population has again fallen to 4,500 people. This trend repeats every year. City B, on the other hand, is a small town not far from City A. Its population has been growing extremely rapidly ever since the arrival of several large, new automobile-assembly plants. There were only 4,000 people living there on January 1, 1990, but its population has grown by 8% every year thereafter.

(a) Find a possible formula for $P_A(t)$, the population of city A in year t. (Assume that $t = 0$ means January 1, 1990.)

(b) Find a formula for $P_B(t)$, the population of city B in year t.

(c) At how many different points in time will the population of town A equal the population of town B?

(d) Approximately when will the population of City B *first* equal the population of City A? When will the population of B *last* equal the population of A?

ANSWER:

(a) The average is 5000 and the amplitude is 500. The period is 1 year and it will have the shape of an inverted cosine curve.

$$P_A = 5000 - 500\cos(2\pi t)$$

(b) $P_B(t) = 4000(1.08)^t$

(c) See Figure 6.31 and count 5 intersections. So the populations are equal five times.

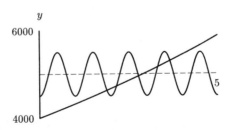

Figure 6.31

(d) Use the trace or intersect feature of a calculator to find $t_1 = 1.89$ and $t_5 = 3.65$. Thus, the populations were first equal towards the end of 1991 (at $t = 1.89$). The populations were last equal slightly after mid 1996 (at $t = 3.65$).

46. The foot of a person riding a stationary bicycle moves in a circle of radius 20 cm centered 30 cm above the ground, making 1 revolution per second. At $t = 0$, the pedal is 10 cm above the ground. Find a formula for $n(t)$, the height of the pedal above the ground at time t (in seconds).

ANSWER:

Since at $t = 0$ the pedal is 10 cm from the ground, we know that at this time, the pedal is at the bottom of its revolution. The height of the pedal is given by the formula

$$n(t) = A\sin(B(t - h)) + k.$$

We know that $k = 30$ since the center of the revolution of the wheel is 30 cm.

We know that $A = 20$ since the radius is 20 cm.

We know that $\frac{2\pi}{B}$ equals the period of the function, which is 1 in our case (since the pedal makes one revolution per second,) so $B = 2\pi$.

We know that h is the right shift of the function. Since at time $t = 0$ we know that the pedal is at the bottom, we know that the sine function is right shifted by one-quarter of a rotation. That is, one-quarter of a second, so

$$h = \frac{1}{4}.$$

Thus

$$n(t) = 20\sin\left(2\pi\left(t - \frac{1}{4}\right)\right) + 30.$$

47. The caribou population in Denali National Park dropped from a high of 200,000 in 1943 to a low of 76,000 in 1989, and has risen some since then. Scientists hypothesize that the population follows a sinusoidal cycle affected by predation and other environmental conditions, and that the caribou population will again reach its previous high.

(a) Letting $t = 0$ in 1943, give a possible sinusoidal formula to describe the caribou population as a function of time.

(b) When does your model predict that the caribou population will next reach 200,000 again?

ANSWER:

(a)

$$P(t) = A\cos(Bt) + k$$

$$A = \text{Amplitude} = \frac{\text{High} - \text{Low}}{2} = \frac{200{,}000 - 76{,}000}{2} = 62{,}000$$

From 1943 to 1989 is one half of a period, so the period is 92 years. This gives

$$\text{Period} = \frac{2\pi}{B} = 92, \quad \text{so} \quad B = \frac{2\pi}{92} = \frac{\pi}{46}.$$

Finally,

$$k = \text{Midline} = \frac{\text{High} + \text{Low}}{2} = \frac{200{,}000 + 76{,}000}{2} = 138{,}000.$$

Thus,

$$P(t) = 62{,}000\cos\left(\frac{\pi}{46}t\right) + 138{,}000$$

(b) The next time the population reaches 200,000 will be one period later, namely $1943 + 92$ years $= 2035$.

48. The graph of $y = f(x)$ is shown in Figure 6.32.

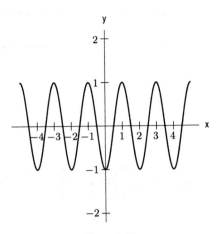

Figure 6.32

(a) Find a formula for f.
(b) Find formulas for the graphs in Figures 6.33 and 6.34 as transformations of f.

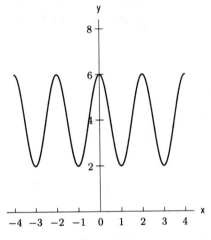

Figure 6.33

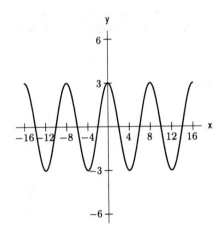

Figure 6.34

ANSWER:

(a) The function f in Figure 6.32 has amplitude=1 and period=2, and since $P = 2 = \dfrac{2\pi}{B}$ we have $B = \pi$. Thus

$$f(x) = \sin(\pi(x - 0.5))$$
$$= \sin(\pi x - \frac{\pi}{2})$$
or
$$f(x) = -\cos(\pi x)$$

(b) For Figure 6.33 we have:

$$y = -2(-\cos(\pi x)) + 4$$
$$y = 2\cos(\pi x) + 4$$
$$y = -2f(x) + 4$$

106

For Figure 6.34 we have period=8, which is 4 times the period of Figure 6.32, and amplitude=3, which is 3 times the amplitude of Figure 6.32. Thus we have:

$$y = -3(-\cos(\frac{\pi x}{4}))$$
$$= 3\cos(\frac{\pi x}{4})$$
$$y = -3f(4x)$$

49. In the physics lab a student collected the data in Table 6.3 on the height of a weight on a spring attached to the ceiling. (The weight was bobbing up and down.)

Table 6.3

Secs	0.0	0.05	0.10	0.15	0.20	0.25	0.30	0.35	0.40	0.45	0.50	0.55
Feet	6.96	7.19	7.40	8.28	8.84	8.45	8.20	7.85	7.23	7.03	7.25	7.85

Find a sinusoidal function that will model this data.
ANSWER:
Looking at the data, we note that at time $t = 0$ the weight is at its minimum height. Thus the function can be modeled by a cosine function which is stretched and displaced vertically but without a phase shift. In particular, since at time $t = 0$ the function will be hitting its minimum, the function must be of the form

$$f(t) = -A\cos(Bt) + k$$

where A is the amplitude, $\frac{2\pi}{B}$ is the period, and k is the midline. In our case, the function varies roughly from 7 to 9 and has a period of about 0.45 seconds. Thus the amplitude is

$$A = \frac{9-7}{2} = 1.$$

The midline is

$$D = \frac{9+7}{2} = 8.$$

And for the period we get

$$0.45 = \frac{2\pi}{B}$$

or simply

$$B = \frac{2\pi}{0.45} \approx 13.96.$$

Thus a function for the height of the weight at time t is

$$f(t) = 8 - \cos(13.96t).$$

50. Write three different formulas for the graph in Figure 6.35.

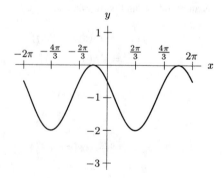

Figure 6.35

ANSWER:
Answers may vary.

$$y = \cos\left(t + \frac{\pi}{3}\right) - 1$$

$$y = -\cos\left(t + \frac{4\pi}{3}\right) - 1$$

$$y = \sin\left(t + \frac{5\pi}{6}\right) - 1$$

51. Graph the function $f(x) = -\sin\left(\frac{x}{2}\right) - 2$ for one period. Specify the amplitude, period, x-intercepts and intervals on which the function is increasing.

ANSWER:
We know that the function

$$A\sin(Bx) + k$$

is a sine function with amplitude —A— vertical shift k and a period of $\frac{2\pi}{B}$. Moreover, if $A < 0$ we know that the function will be an inverted sine function. In our case

$$A = -1$$
$$B = \frac{1}{2}$$

and

$$k = -2.$$

Thus, we know that the amplitude of the function is 1. The period of the function is

$$\frac{2\pi}{B} = \frac{2\pi}{1/2} = 4\pi.$$

The x-intercept is the value of x for which $f(x) = 0$. Since we know that the function has been vertically shifted by -2 and that the amplitude of the function is 1, we know that the maximum of the function is

$$\text{max of } f(x) = -2 + |A| = -2 + 1 = -1$$

and the minimum is

$$\text{min of } f(x) = -2 - |A| = -2 - 1 = -3$$

thus there is no value of x for which $f(x) = 0$. We know that the function $y = \sin x$ is decreasing on the interval $\frac{\pi}{2} < x < \frac{3\pi}{2}$. Thus we know that the function $y = \sin\frac{x}{2} - 2$ is decreasing on the interval $\pi < x < 3\pi$. So the function $y = -\sin\frac{x}{2} - 2$ is increasing on the interval $\pi < x < 3\pi$.

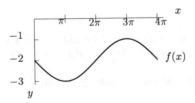

Figure 6.36

52. A vacationer sits all day on the corner of a pier in Boston Harbor and notices that at 9 am, when the water level is at its lowest, the water's depth is 2 feet. At 4 pm, the water has risen to its maximum depth of 12 feet. If the depth of the water level varies sinusoidally, find a formula for the depth of the water as a function of time, t, since 9 am. Sketch a graph of your function.

ANSWER:
If we set up our function as $f(t)$ where t is the number of hours after 9 am, we know that the function will resemble an inverted cosine curve (since it will hit its minumum at time $t = 0$.) Thus a formula for the function $f(x)$ will be of the form

$$f(x) = -A\cos(Bt) + k$$

108

where A is the amplitude, $\frac{2\pi}{B}$ is the period and k is the midline. Since we know that the maximum depth is 12 feet and the minimum depth is 2 feet we know that

$$\text{The amplitude, } A = \frac{\text{max depth} - \text{min depth}}{2} = \frac{12-2}{5} = \frac{10}{2} = 5$$

and

$$\text{The midline, } k = \frac{\text{max depth} + \text{min depth}}{2} = \frac{12+2}{5} = \frac{14}{2} = 7.$$

We also know that the tide goes from a minimum depth to a maximum depth in 7 hours, that is it has a half-period of 7 or a full period of 14 hours so

$$\frac{2\pi}{B} = 14$$

or in other words

$$B = \frac{\pi}{7}.$$

Thus we get

$$f(t) = -5\cos\left(\frac{\pi}{7}t\right) + 7.$$

A graph of this function is shown in Fig 6.37.

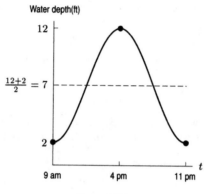

Figure 6.37

53. The position, S, of a piston in a 6-inch stroke in an engine is given as a function of time, t, in seconds, by the formula

$$S = 3\sin(200\pi t).$$

(a) What is the period of this function?
(b) Sketch a graph of one period of this function. Label the units on your axes clearly.
(c) Why is this engine said to have a 6-inch stroke? (In other words, what feature of the graph corresponds to the 6 inches?)
(d) How many revolutions per minute (rpm) is this engine performing?

ANSWER:

(a) The period of the motion is given by

$$P = \frac{2\pi}{200\pi} = \frac{1}{100}\sec = 0.01\sec$$

(b) The graph of S is in Figure 6.38.

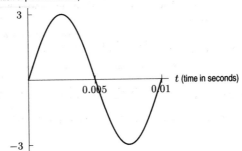

S (position of piston in inches)

Figure 6.38

(c) The total displacement of a full stroke is 6 in. This can be seen in the graph where the distance between the highest point ($+3$ in) and the lowest point (-3 in) is 6 in.

(d) Since the period is 0.01 sec, and 1 minute $= 60$ seconds, the piston is performing at a rate of

$$\frac{60}{0.01} = 6000 \text{ revolutions per minute.}$$

54. The point Q in Figure 6.39 is at the lowest point on the sine-curve. As k increases, does the value of m

(a) Increase
(b) Decrease
(c) Stay the same
(d) Can't tell

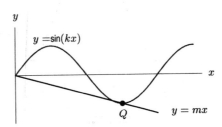

Figure 6.39

ANSWER:
As k increases, the period of the sine curve decreases, moving point Q closer to the origin. Thus, the slope of the line becomes more negative and the value of m decreases.

55.

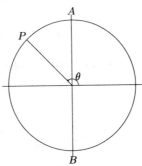

Figure 6.40

As P moves counterclockwise around the circle from A to B, does

(a) $\sin \theta$ increase, decrease, increase then decrease, or decrease and then increase?
(b) $\cos \theta$ increase, decrease, increase then decrease, or decrease and then increase?
(c) $\tan \theta$ increase, decrease or stay the same?

ANSWER:

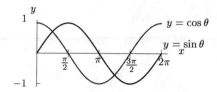

Figure 6.41

Note that as P moves counterclockwise around the circle from A to B, θ traverses the angles from $\frac{\pi}{2}$ to $\frac{3\pi}{2}$.

(a) Looking at Figure 6.41 we see that on the interval $\frac{\pi}{2} \le \theta \le \frac{3\pi}{2}$ the sine function is decreasing.

(b) Looking at Figure 6.41 we see that on the interval $\frac{\pi}{2} \le \theta \le \frac{3\pi}{2}$ the cosine function first decreases and then increases.

(c)

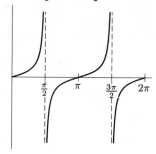

Figure 6.42

Looking at the graph of the tangent function in Figure 6.42 we see that $\tan \theta$ increases on the interval $\frac{\pi}{2} < \theta < \frac{3\pi}{2}$.

56. Next to each expression below, write the line segment in the figure whose length is equal to the value of the expression. (Notice that the circle has radius 1, so write fractions with a numerator or a denominator equal to 1.

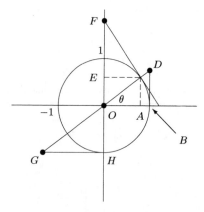

Figure 6.43

(a) $\sin \theta$ = length of segment _____
(b) $\cos \theta$ = length of segment _____
(c) $\tan \theta$ = length of segment _____
(d) $\csc \theta$ = length of segment _____
(e) $\sec \theta$ = length of segment _____
(f) $\cot \theta$ = length of segment _____

ANSWER:

(a) OE **(b)** OA **(c)** BD **(d)** OG **(e)** OD **(f)** GH

57. Match the function (a) to (d) with a quotient I to IV:

   (a) $\tan t$ **(b)** $\cot t$ **(c)** $\csc t$ **(d)** $\sec t$

  I. $\dfrac{1}{\tan t}$ II. $\dfrac{1}{\cos t}$ III. $\dfrac{1}{\sin t}$ IV. $\dfrac{\sin t}{\cos t}$

  ANSWER:

  (a) IV
  (b) I
  (c) III
  (d) II

58. Are the following statements true or false? Give a reason for your answer.

  (a) The period of $f(x) = \cos(x/3)$ is greater than the period of $g(x) = \tan(x/3)$.
  (b) The graphs of $f(x) = \tan x$ and $g(x) = x^2$ cross infinitely often.
  (c) The graphs of $f(x) = \cos x$ and $g(x) = x^2$ cross infinitely often.

  ANSWER:

  (a) True. The period of $f(x) = \cos(x/3)$ is $3 \cdot 2\pi = 6\pi$; the period of $g(x) = \tan(x/3)$ is 3π.
  (b) True. The graph of $g(x)$ will cross the graph of $f(x)$ once between every asymptote of $f(x)$, because $f(x)$ ranges from $-\infty$ to ∞ in that interval.
  (c) False. Outside the interval $-1 \leq x \leq 1$, the values of $g(x)$ are above 1, and so the graphs cannot intersect for $x > 1$ or for $x < -1$. Inside the interval $-1 \leq x \leq 1$, the graphs cross twice.

59. Let $f(x) = \sin^{-1} x$, $g(x) = \cos^{-1} x$ and $h(x) = f(x) + g(x)$, where all the angles are measured in radians,

  (a) Graph all three functions on the same graph, labeling all points of intersection and at least four other points on the graph.
  (b) What is the domain of $h(x)$?
  (c) Using the insight gained from the graph, simplify the formula for $h(x)$. Explain why this makes sense.

  ANSWER:

  (a) The three intersection points are $(0, \pi/2)$, $(1, \pi/2)$, $and (\sqrt{2}/2, \pi/4)$.

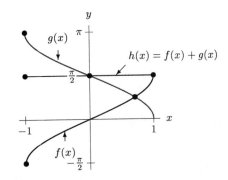

Figure 6.44

  (b) We know that the function $f(x)$ and $g(x)$ are both defined on the interval $-1 \leq x \leq 1$ so the function $h(x) = f(x) + g(x)$ is defined on the interval

$$-1 \leq x \leq 1$$

  (c) From the graph we see that

$$h(x) = \sin^{-1} x + \cos^{-1} x = \frac{\pi}{2}.$$

If $\sin^{-1} x = \alpha$, then $\sin \alpha = x$. If $\cos^{-1} x = \beta$ then $\cos \beta = x$. Thus the formula states that if $\sin \alpha = \cos \beta$ then $\alpha + \beta = \frac{\pi}{2}$, that is, α and β are complementary angles. This fact can be shown in a right triangle:

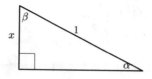

Figure 6.45

In this triangle, $\alpha + \beta = \frac{\pi}{2}$, $\sin \alpha = \frac{x}{1} = x$ and $\cos \beta = \frac{x}{1} = x$. Thus we see that in fact, when $\alpha + \beta = \frac{\pi}{2}$ then $\sin \alpha = \cos \beta$.

60. Solve exactly for x:

(a) $\cos x = \frac{-\sqrt{3}}{2}$

(b) $\cos^{-1}\left(\frac{-\sqrt{3}}{2}\right) = x$

ANSWER:

(a) We are asked to find the angle x with cosine $\frac{-\sqrt{3}}{2}$. We know that

$$\cos \frac{\pi}{6} = \frac{\sqrt{3}}{2},$$

thus we know that

$$\cos\left(\pi - \frac{\pi}{6}\right) = \cos \frac{5\pi}{6} = \frac{-\sqrt{3}}{2}.$$

We also know that $\frac{7\pi}{6}$ has the same cosine value as $\frac{5\pi}{6}$. Since the cosine function has a period of 2π, all solutions are of the form

$$x = \frac{5\pi}{6} + 2k\pi \quad \text{or} \quad x = \frac{7\pi}{6} + 2k\pi$$

where k is any integer.

(b) We are looking for the x such that

$$\cos^{-1} \frac{-\sqrt{3}}{2} = x$$

or, in other words,

$$\cos x = \frac{-\sqrt{3}}{2}.$$

We know that the range of the $\cos^{-1}$ function is $-\frac{\pi}{2} \le x \le \frac{\pi}{2}$. Thus we are looking for the solution from part (a) which is within this range. The only solution is

$$x = \frac{5\pi}{6}.$$

61. For each equation, find all exact solutions for $0 \le \theta \le 2\pi$.

(a) $\tan \theta = -\sqrt{3}$ (b) $\sin^2 \theta = \frac{1}{4}$ (c) $\cos \theta = \frac{1}{5}$

ANSWER:

(a) $\tan \theta = -\sqrt{3}$, so $\theta = \tan^{-1}(-\sqrt{3})$ For $0 < \theta < 2\pi, \theta$ will be in the second and fourth quadrants:

$$\theta = \frac{2\pi}{3}, \frac{5\pi}{3}.$$

(b) $\sin^2 \theta = \frac{1}{4}$ so $\sin \theta = \pm\frac{1}{2}$.

$$\theta = \sin^{-1}\left(-\frac{1}{2}\right), \sin^{-1}\left(\frac{1}{2}\right)$$

For $\theta = \sin^{-1}\left(-\frac{1}{2}\right), \theta = \frac{7\pi}{6}, \frac{11\pi}{6}$. For $\theta = \sin^{-1}\left(\frac{1}{2}\right), \theta = \frac{\pi}{6}, \frac{5\pi}{6}$. Values of θ are: $\frac{\pi}{6}, \frac{5\pi}{6}, \frac{7\pi}{6}, \frac{11\pi}{6}$.

(c) $\cos \theta = \frac{1}{5}$, so $\theta = \cos^{-1}(\frac{1}{5})$. For $0 \le \theta \le 2\pi, \theta$ will lie in the first and fourth quadrants where the cosine function assumes positive values:

$$\theta = 1.37, 4.91.$$

62. Evaluate exactly. (Note: *None* of the expressions simplifies to 4.)

    (i) $\sin^{-1}(\sin 4)$ (ii) $\cos^{-1}(\cos 4)$ (iii) $\tan^{-1}(\tan 4)$ (iv) $\sin(\sin^{-1}(4))$

    ANSWER:

    (a) $\sin 4 = -0.7568$,   so $\sin^{-1}(\sin 4) = \sin^{-1}(-0.7568) = -0.8584$.

    (b) $\cos 4 = -0.6536$,   so $\cos^{-1}(\cos 4) = \cos^{-1}(-0.6536) = 2.2832$.

    (c) $\tan 4 = 1.1578$,   so $\tan^{-1}(\tan 4) = \tan^{-1}(1.1578) = 0.8584$.

    (d) $\sin^{-1}(4)$ is not defined on the set of real numbers since $-1 \leq \sin \theta \leq 1$.

63. **(a)** Graph $g(x) = \sin^{-1}(2\pi x)$.

    (b) Give the domain of $g(x)$.

    (c) Give the range of $g(x)$.

    ANSWER:

(a)

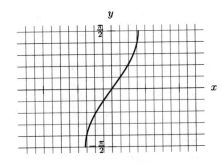

Figure 6.46

    (b) We know that $\sin^{-1}(x)$ has a domain of -1 to 1. Thus $g(x) = \sin^{-1}(2\pi x)$ must have a domain of

$$-\frac{1}{2\pi} \leq x \leq \frac{1}{2\pi}.$$

    (c) Like the function $\sin^{-1}(x)$, $g(x)$ has a range of

$$-\frac{\pi}{2} \leq g(x) \leq \frac{\pi}{2}.$$

64. The Big Ben in London is 320 feet high. How long will the new London ferris wheel rider be above the Big Ben, during a ride of one revolution?

    (a) as a rough answer? (Accurate to the minute.)

    (b) as an engineering answer? (Accurate to the second.)

    (c) as a mathematical answer? (*Exact* expression.)

    ANSWER:

    We know that the expression for the height of a rider on the new London ferris wheel is

$$h(t) = 250 - 250 \cos\left(\frac{\pi}{10}t\right).$$

Thus we are asked to solve for t such that

$$h(t) \geq 320.$$

Solving for t such that $h(t) = 320$ we get

$$320 = h(t)$$
$$= 250 - 250 \cos\left(\frac{\pi}{10}t\right)$$
$$70 = -250 \cos\left(\frac{\pi}{10}t\right)$$
$$-\frac{70}{250} = \cos\left(\frac{\pi}{10}t\right)$$
$$\arccos\left(-\frac{7}{25}\right) = \frac{\pi}{10}t$$
$$t = \frac{10}{\pi} \arccos\left(-\frac{7}{25}\right)$$
$$\approx 5.9033$$

Since we know that the period of $h(t)$ is 20 minutes we know that another solution for $h(t) = 320$ is

$$t = 20 - \frac{10}{\pi} \arccos\left(-\frac{7}{25}\right) \approx 14.0967.$$

In between these two values of t the rider is higher than 320 feet. Thus we get:

(a) $14.0967 - 5.9033 = 8.1934$ and so as a rough answer the rider is above Big Ben for 8 minutes.

(b) We know that one minute is 60 seconds, so that 0.1934 minutes is

$$60 * 0.1934 \approx 11.604 \text{ seconds.}$$

Thus, the engineering answer is that the rider is above Big Ben for 8 minutes and 12 seconds.

(c) The exact mathematical answer is:

$$20 - \frac{10}{\pi} \arccos\left(-\frac{7}{25}\right) - \frac{10}{\pi} \arccos\left(-\frac{7}{25}\right)$$

or

$$\frac{10}{\pi}\left(2\pi - 2\arccos\left(-\frac{7}{25}\right)\right).$$

65. Suppose that α is in the first quadrant and that $\beta = \pi + \alpha$. What is $\cos^{-1}(\cos\beta)$?

ANSWER:

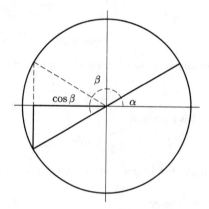

Figure 6.47

See Figure 6.47. We know that $\cos^{-1}(\cos\beta)$ is the angle between 0 and π whose cosine is the same as the cosine of β. Therefore,

$$\cos^{-1}(\cos\beta) = \pi - \alpha.$$

Chapter 7 Exam Questions

Exercises

1. Use Figure 7.1 to find the missing side, c, and the missing angles, A and B (if possible), when $a = 8$, $b = 7$, and $C = 50°$.

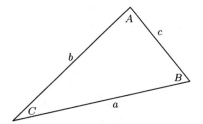

Figure 7.1

ANSWER:

Using the Law of Cosines,

$$c^2 = 8^2 + 7^2 - 2(8)(7)\cos 50°$$
$$= 113 - 112(0.6427876)$$
$$= 41.0077888$$

Thus, $c = \sqrt{41.0077888} = 6.4037$.

To find A, use the Law of Sines:

$$\frac{\sin A}{8} = \frac{\sin 50°}{6.4037}$$
$$\sin A = 0.9570$$
$$A = \sin^{-1} 0.9570 = 73.14°.$$

Similarly,

$$\frac{\sin B}{7} = \frac{\sin 50°}{6.4037}$$
$$\sin B = 0.83738$$
$$B = \sin^{-1} 0.83738 = 56.86°.$$

2. Use Figure 7.2 to find the missing sides, a and b, and the missing angle, C, when $A = 28°$, $B = 45°$, and $c = 12$.

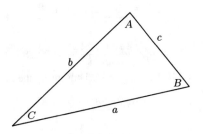

Figure 7.2

ANSWER:
We know all three angles add up to $180°$, so $C = 180 - (28 + 45) = 107°$. To find a, use the Law of Sines:

$$\frac{\sin 28°}{a} = \frac{\sin 107°}{12}$$

$$a = \frac{\sin 28°(12)}{\sin 107°} = 5.89.$$

Similarly,

$$\frac{\sin 45°}{b} = \frac{\sin 107°}{12}$$

$$b = \frac{\sin 45°(12)}{\sin 107°} = 8.87.$$

3. Use Figure 7.3 to find the missing side, a, and the missing angles, B and C (if possible), when $b = 9$, $c = 11$, and $B = 98°$.

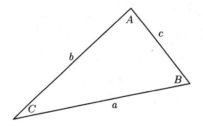

Figure 7.3

ANSWER:
Using the Law of Sines:

$$\frac{\sin C}{11} = \frac{\sin 98°}{9}$$

$$\sin C = \frac{\sin 98°(11)}{9} = 1.2103$$

But this is impossible, since 1.2103 is not in the range of the Sine function. Thus, there is no such triangle possible.

4. Find all sides and angles of a triangle with $b = 27$ cm, $c = 14$ cm, and $\gamma = 28°$. Sketch the triangle (if there is more than one possible triangle, solve and sketch both). Note that α is the angle opposite side a, β is the angle opposite side b, and γ is the angle opposite side c.
ANSWER:
Using the Law of Sines:

$$\frac{\sin B}{27} = \frac{\sin 28°}{14}$$

$$\sin B = \frac{27 \sin 28°}{14} = 0.9054$$

$$B = 64.88° \text{ or } B = 180 - 64.88 = 115.12°.$$

For the first case, $A = 180 - (28 + 64.88) = 87.12°$. Thus, using the Law of Sines again:

$$\frac{\sin 87.12°}{a} = \frac{\sin 28°}{14}$$

$$a = \frac{14 \sin 87.12°}{\sin 28°} = 29.78.$$

For the second case, $A = 180 - (28 + 115.12) = 36.88°$. Thus, using the Law of Sines again:

$$\frac{\sin 36.88°}{a} = \frac{\sin 28°}{14}$$

$$a = \frac{14 \sin 36.88°}{\sin 28°} = 17.90.$$

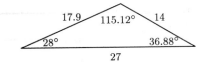

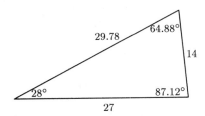

Figure 7.4

5. Complete the following table, using exact values where possible.

Table 7.1

θ(radians)	$\sin^2\theta$	$\cos^2\theta$	$\sin 2\theta$	$\cos 2\theta$
0				
$\pi/4$				
π				

ANSWER:

Table 7.2

θ(radians)	$\sin^2\theta$	$\cos^2\theta$	$\sin 2\theta$	$\cos 2\theta$
0	0	1	0	1
$\pi/4$	1/2	1/2	1	0
π	0	1	0	1

6. Simplify the expression $\dfrac{\cos 2\theta}{\cos\theta - \sin\theta}$.

ANSWER:

$$\frac{\cos 2\theta}{\cos\theta - \sin\theta} = \frac{\cos^2\theta - \sin^2\theta}{\cos\theta - \sin\theta}$$
$$= \frac{(\cos\theta - \sin\theta)(\cos\theta + \sin\theta)}{\cos\theta - \sin\theta}$$
$$= \cos\theta + \sin\theta$$

7. Simplify the expression $\dfrac{1}{\cos\theta + \sin\theta} + \dfrac{1}{\cos\theta - \sin\theta}$

ANSWER:

$$\frac{1}{\cos\theta + \sin\theta} + \frac{1}{\cos\theta - \sin\theta} = \frac{(\cos\theta - \sin\theta) + (\cos\theta + \sin\theta)}{(\cos\theta + \sin\theta)(\cos\theta - \sin\theta)}$$
$$= \frac{2\cos\theta}{\cos^2\theta - \sin^2\theta}$$

8. Write $4\sin t + 2\cos t$ in the form $A\sin(Bt + \phi)$ using sum or difference formulas.

ANSWER:

We get $A = \sqrt{4^2 + 2^2} = 2\sqrt{5}$, and $\tan\phi = 2/4 = 1/2$. Since $\cos\phi = 4/(2\sqrt{5})$ and $\sin\phi = 2/(2\sqrt{5})$ are both positive, ϕ is in the first quadrant, and $\phi = \tan^{-1}(1/2) = 0.4636$. Thus,

$$4\sin t + 2\cos t = 2\sqrt{5}\sin(t + 0.4636).$$

9. Write $3\sin(2t) - 4\cos(2t)$ in the form $A\sin(Bt + \phi)$ using sum or difference formulas.

ANSWER:

We get $A = \sqrt{3^2 + 4^2} = 5$, and $\tan\phi = -4/3$. Since $\cos\phi = 3/5$ is positive and $\sin\phi = -4/5$ is negative, ϕ is in the fourth quadrant, and $\phi = \tan^{-1}(-4/3) = -0.9273$. Thus,

$$3\sin(2t) - 4\cos(2t) = 5\sin(2t - 0.9273)$$

10. Without a calculator, find the exact value of $\sin 165°$.

ANSWER:

Using sum- and difference-of-angle formulas,

$$\begin{aligned}
\sin 165° &= \sin(120° + 45°) \\
&= \sin 120° \cos 45° + \cos 120° \sin 45° \\
&= \frac{\sqrt{3}}{2}\frac{\sqrt{2}}{2} - \frac{1}{2}\frac{\sqrt{2}}{2} \\
&= \frac{\sqrt{2}(\sqrt{3} - 1)}{4}
\end{aligned}$$

11. Without a calculator, find the exact value of $\cos(-75°)$.

ANSWER:

First, $\cos(-75°) = \cos 75°$. Then, using sum- and difference-of-angle formulas,

$$\begin{aligned}
\cos 75° &= \cos(30° + 45°) \\
&= \cos 30° \cos 45° - \sin 30° \sin 45° \\
&= \frac{\sqrt{3}}{2}\frac{\sqrt{2}}{2} - \frac{1}{2}\frac{\sqrt{2}}{2} \\
&= \frac{\sqrt{2}(\sqrt{3} - 1)}{4}
\end{aligned}$$

12. Graph the function $f(x) = 3e^{-0.23x}\sin x$ on the interval $0 < x < 15$. Use a calculator or computer to find the maximum and minimum values of the function and where they occur.

ANSWER:

$f(x)$ has a maximum of 2.15 at $x = 1.34$ and a minimum of -1.04 at $x = 4.49$.

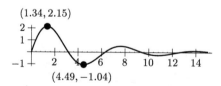

Figure 7.5

13. Sketch and describe the graph of $y = \cos x + \cos 2x$ on $0 \le x \le 4\pi$.

ANSWER:

y is periodic with period 2π.

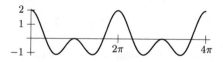

Figure 7.6

14. Graph $y = 10 + \frac{1}{2}t$ and $y = 10 + \frac{1}{2}t - 4\cos\left(\frac{t}{2}\right)$ on $0 \le t \le 6\pi$. Where would the graphs intersect for larger values of t?

ANSWER:

The graphs intersect at odd multiples of π.

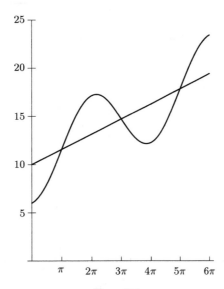

Figure 7.7

15. In which quadrant is a point with the polar coordinate θ?

    (a) $100°$ **(b)** $-512°$ **(c)** 3.7π **(d)** $-7\pi/4$

    ANSWER:

    (a) II   **(b)** III   **(c)** IV   **(d)** I

16. Find polar coordinates of the points with the following Cartesian coordinates.

    (a) $(-5, 0)$ **(b)** $(2, 2)$ **(c)** $(0, -7\pi)$ **(d)** $(-5, -5)$

    ANSWER:

    (a) $r = 5, \theta = \pi$
    (b) $r = \sqrt{2^2 + 2^2} = \sqrt{8}, \theta = \pi/4$
    (c) $r = 7\pi, \theta = 3\pi/2$
    (d) $r = \sqrt{(-5)^2 + (-5)^2} = \sqrt{50}, \theta = 5\pi/4$

17. What are the Cartesian coordinates of the points with the following polar coordiantes?

    (a) $(1, -\pi/4)$ **(b)** $(0, \pi)$ **(c)** $(\sqrt{2}, 3\pi/4)$ **(d)** $(3, \pi/3)$

    ANSWER:

    (a) $x = \cos(-\pi/4) = 1/\sqrt{2}$ and $y = \sin(-\pi/4) = -1/\sqrt{2}$.
    (b) $x = y = 0$
    (c) $x = \sqrt{2}\cos(3\pi/4) = \sqrt{2}(-1/\sqrt{2}) = -1$ and $y = \sqrt{2}\sin(3\pi/4) = \sqrt{2}(1/\sqrt{2}) = 1$.
    (d) $x = 3\cos(\pi/3) = 3/2$ and $y = 3\sin(\pi/3) = 3\sqrt{3}/2$.

18. Express the complex number $3i$ in polar form, $z = re^{i\theta}$.

    ANSWER:

    For $z = 3i$, the distance from z to the origin is 3, so $r = 3$. Also, the angle is $\theta = \pi/2$. Thus, $z = 3e^{i\pi/2}$.

19. Express the complex number $1 - 4i$ in polar form, $z = re^{i\theta}$.

    ANSWER:

    For $z = 1 - 4i$, the distance from z to the origin is $\sqrt{1^2 + (-4)^2} = \sqrt{17} = 4.123$, so $r = 4.123$. Also, $\tan\theta = -4$ so, for example, $\theta = -1.326$. Thus, $z = 4.123e^{-1.326i}$.

20. Perform the calculation $(2 + i)(3 - 2i)$ and give your answer in Cartesian form, $z = x + iy$.
ANSWER:

$$(2 + i)(3 - 2i) = 6 - 4i + 3i - 2i^2$$
$$= 8 - i$$

21. Perform the calculation $\sqrt{9e^{i\pi/3}}$ and give your answer in Cartesian form, $z = x + iy$.
ANSWER:

$$\sqrt{9e^{i\pi/3}} = 3e^{i\pi/6}$$
$$= 3(\cos \pi/6 + i \sin \pi/6)$$
$$= \frac{3\sqrt{3}}{2} + \frac{3}{2}i$$

Problems

22. Let θ and ϕ be the two angles shown in Figure 7.8 and suppose $0 < \theta < 90°$ and $90° < \phi < 180°$. Note: Side lengths are not to scale.

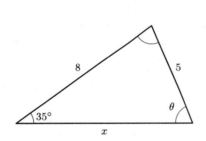

Figure 7.8

(a) Find θ and ϕ.
(b) Find x and y.
ANSWER:

(a) The Law of Sines tells us that in a triangle,
$$\frac{\sin A}{a} = \frac{\sin B}{b}$$
where a is the length of the side opposite the angle A and b is the length of the side opposite angle B. Thus

$$\frac{\sin \theta}{8} = \frac{\sin 35}{5}$$
$$\sin \theta = \frac{8 \sin 35}{5}$$
$$\theta = \sin^{-1}\left(\frac{8 \sin 35}{5}\right)$$
$$\approx 66.6°.$$

On the other hand, since we know that $\sin x = \sin(180 - x)$ we know

$$\phi \approx 180 - 66.6 \approx 113.4°.$$

(b) Given the length of two sides a and b and the angle between them C, we know, by the Law of Cosines, that the length of the third side c satisfies
$$c^2 = a^2 + b^2 - 2ab \cos C.$$

In the first triangle we let $a = 8, b = 5, c = x$ and $C = 180 - 35 - 66.6 = 78.4°$.
Solving we get

$$
\begin{aligned}
x^2 &= 64 + 25 - 2(8)(5)\cos 78.4 \\
&= 89 - 80\cos 78.4 \\
&\approx 89 - 80(.2) \\
&= 89 - 16 \\
&= 73 \\
x &= \sqrt{73} \\
&\approx 8.5.
\end{aligned}
$$

In the second triangle we let $a = 8, b = 5, c = y$, and $C = 180 - 35 - 113.4 = 31.6°$.
Solving we get

$$
\begin{aligned}
y^2 &= 64 + 25 - 2(8)(5)\cos 31.6 \\
&= 89 - 80\cos 31.6 \\
&\approx 89 - 80(.85) \\
&= 89 - 68 \\
&= 21 \\
y &= \sqrt{21} \\
&\approx 4.6.
\end{aligned}
$$

23. The boundaries of the Bermuda Triangle are not universally agreed upon, but one version places one corner of the triangle on the southern US coast (point A in Figure 7.9), another corner in the Bermuda Islands (point B), and the third corner in the Greater Antilles—Cuba, Jamaica, Puerto Rico and Hispaniola (point C).

Assume that the distance between A and B is 1000 km, and the angle at point A is 55°. Traveling at 25 km/hr, a cruise ship takes 26 hours to cross from point A to point C. If the ship survives this portion of the trip, what is the distance it must travel from point C to point B?

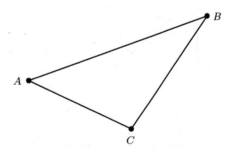

Figure 7.9: The Bermuda Triangle

ANSWER:
We have the length of side $AB = 1000$ and the length of side $AC = 25$ km/hr $\cdot$ 26 hr $= 650$.

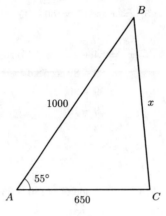

Figure 7.10

From Figure 7.10, we see that if x is the length of side BC, then by the Law of Cosines,

$$x^2 = 650^2 + 1000^2 - 2 \cdot 650 \cdot 1000 \cdot \cos(55°)$$
$$x^2 = 676{,}850$$
$$x = 822.7.$$

This ship must travel 822.7 km.

24. In order to measure the distance across the lake from A to B, a surveyor is able to measure from a third point, C, on land. The distance AC is 100 meters and BC is 120 meters. He finds that angle ACB is 47°. Find distance AB. (The figure is not drawn to scale.)

   ANSWER:

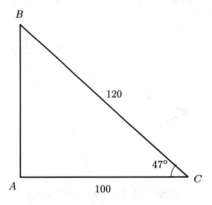

Figure 7.11

Using the Law of Cosines, we get:

$$AB^2 = 100^2 + 120^2 - 2(100)(120) \cos 47°$$

$$AB \approx 89.6 \text{ meters}$$

25.

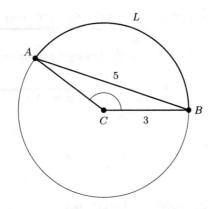

Figure 7.12

Find the length of arc L in Figure 7.12.

ANSWER:

In $\triangle CAB$, $AB = 5$, $CB = 3$. Also, $CA = 3$, since $CA = CB = 3 = $ Radius. Using the Law of Cosines on $\triangle CAB$ gives us:

$$5^2 = 3^2 + 3^2 - 2 \cdot 3 \cdot 3 \cdot \cos C$$
$$25 = 9 + 9 - 18 \cos C$$
$$\cos C = -\frac{7}{18}.$$

So,

$$C = \cos^{-1}\left(-\frac{7}{18}\right) = 1.97 \text{ radians}$$

Now, we know that $L = r\theta$, provided that θ is in radians, so

$$l = 3 \cdot 1.97 = 5.91.$$

26. Solve for θ, with $-\pi \leq \theta \leq \pi$:

$$2 \sin \theta \left(\cos \theta + \frac{1}{\sin \theta}\right) = 1.3$$

ANSWER:

Solving we get

$$1.3 = 2 \sin \theta \left(\cos \theta + \frac{1}{\sin \theta}\right)$$
$$= 2 \sin \theta \cos \theta + 2$$
$$-0.7 = 2 \sin \theta \cos \theta$$

We now recall the double angle formula that tells us that

$$\sin 2\theta = 2 \sin \theta \cos \theta.$$

Thus we have

$$\sin 2\theta = -0.7$$
$$2\theta = -0.78 + 2k\pi$$
$$\theta = -0.39 + k\pi$$

where $k = 0, \pm 1, \pm 2, \cdots$. The value $2\theta = -0.78$ means that 2θ is in Quadrant IV. However, 2θ could be in Quadrant III with $2\theta = \pi + 0.78$. Thus, we can also have

$$\theta = 1.96 + k\pi$$

But we are asked to find all solutions of θ for $-\pi \leq \theta \leq \pi$. Thus, the four solutions are,

$$\theta_1 = -0.39$$
$$\theta_2 = -0.39 + \pi \approx 2.75$$
$$\theta_3 = 1.96$$
$$\theta_4 = 1.96 - \pi \approx -1.18$$

27. Solve for x *exactly* for $0 \leq x \leq 2\pi$:

$$3\sin^2 x - \cos^2 x = 2.$$

ANSWER:
We can substitute $\cos^2 x$ with $1 - \sin^2 x$ getting the equation

$$3\sin^2 x - (1 - \sin^2 x) = 2$$
$$4\sin^2 x = 3$$
$$\sin x = \pm\frac{\sqrt{3}}{2}$$

Solving for x with $0 \leq x \leq 2\pi$ we get

$$x = \frac{\pi}{3}, \; \frac{2\pi}{3}, \; \frac{4\pi}{3}, \; \frac{5\pi}{3}.$$

28. Solve for x for $-\pi \leq x \leq 2\pi$:

$$\sin^3 x + \sin x \cos^2 x = \cos x$$

ANSWER:
We have

$$\sin x \underbrace{\left(\sin^2 x + \cos^2 x\right)}_{1} = \cos x.$$

So in other words

$$\sin x = \cos x.$$

We know that the solutions to the equation $\sin x = \cos x$ are $\frac{\pi}{4} + k\pi$ where k is an integer. Thus, on the interval $-\pi \leq x \leq 2\pi$, the solutions are

$$x = -\frac{3\pi}{4}, \quad x = \frac{\pi}{4}, \quad x = \frac{5\pi}{4}.$$

29. Simplify as much as possible:

$$\frac{(\cos\theta + \sin\theta)^2 - (\cos\theta - \sin\theta)^2}{2\sin 2\theta}.$$

ANSWER:
We have, on expanding parentheses in the numerator,

$$\frac{(\cos\theta + \sin\theta)^2 - (\cos\theta - \sin\theta)^2}{2\sin 2\theta} = \frac{\cos^2\theta + 2\cos\theta\sin\theta + \sin^2\theta - (\cos^2\theta + \sin^2\theta - 2\cos\theta\sin\theta)}{2\sin 2\theta}$$

$$= \frac{4\cos\theta\sin\theta}{2\sin 2\theta}$$
$$= \frac{2\cos\theta\sin\theta}{\sin 2\theta}$$
$$= 1 \qquad \text{Using } \sin 2\theta = 2\cos\theta\sin\theta$$

30. Solve exactly for θ, with $0 \le \theta \le 2\pi$:

$$\cos^4 \theta - \sin^4 \theta = \frac{1}{2}.$$

ANSWER:

We have $\cos^4 \theta - \sin^4 \theta = \frac{1}{2}$ which factors to

$$(\cos^2 \theta - \sin^2 \theta)(\cos^2 \theta + \sin^2 \theta) = \frac{1}{2}.$$

Thus

$$(\cos(2\theta))(1) = \frac{1}{2} \quad \text{so} \quad \cos(2\theta) = \frac{1}{2}.$$

One solution is

$$2\theta = \cos^{-1}\left(\frac{1}{2}\right) = \frac{\pi}{3}.$$

If $0 \le \theta \le 2\pi$, then $0 \le 2\theta \le 4\pi$. Other angles with $\cos(2\theta) = \frac{1}{2}$ are $2\theta = \frac{\pi}{3}, \frac{5\pi}{3}, \frac{7\pi}{3}, \frac{11\pi}{3}$, so $\theta = \frac{\pi}{6}, \frac{5\pi}{6}, \frac{7\pi}{6}, \frac{11\pi}{6}$.

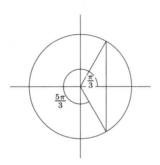

Figure 7.13

31. Let $f(x) = 3\sin(\pi x)$.

   (a) What is the period of $f(x)$?
   (b) What is the amplitude of $f(x)$?
   (c) Write $f(x)$ as a cosine function.

ANSWER:

In a function of the form $f(x) = A\sin(Bx)$, we know that A is the amplitude and $\frac{2\pi}{B}$ is the period. Thus in our case we get:

   (a) The period of $f(x)$ is

$$\frac{2\pi}{\pi} = 2.$$

   (b) The amplitude of $f(x)$ is 3.
   (c) We know that

$$\sin(x) = \cos\left(x - \frac{\pi}{2}\right).$$

Thus we get

$$A\sin(x) = A\cos\left(x - \frac{\pi}{2}\right).$$

So replacing A with 3 and x with πx we get

$$3\sin(\pi x) = 3\cos\left(\pi x - \frac{\pi}{2}\right).$$

32. Simplify completely: $\dfrac{\tan^2 \theta \cos \theta}{(1 - \cos \theta)(1 + \cos \theta)}$

ANSWER:

Using $\tan \theta = (\sin \theta)/(\cos \theta)$, we have

$$\frac{\tan^2 \theta \cos \theta}{(1 - \cos \theta)(1 + \cos \theta)} = \frac{\sin^2 \theta}{\cos^2 \theta} \cdot \frac{\cos \theta}{1 - \cos^2 \theta} = \frac{\sin^2 \theta}{\cos^2 \theta} \cdot \frac{\cos \theta}{\sin^2 \theta} = \frac{1}{\cos \theta}.$$

33. (a) Write the sum function $s(x) = \sin(2x) + \sin(x)$ as a product of sines and cosines.

(b) Show how to use the Pythagorean identity to derive the identity:

$$1 + \tan^2(x) = \sec^2(x).$$

ANSWER:

(a) We have

$$\sin(a+b) + \sin(a-b) = \sin(a)\cos(b) + \sin(b)\cos(a) + \sin(a)\cos(b) - \sin(b)\cos(a)$$
$$= 2\sin(a)\cos(b)$$

In our case, setting $a + b = 2x$ and $a - b = x$ we get $a = \frac{3x}{2}$ and $b = \frac{x}{2}$. Thus we get

$$s(x) = 2\sin\left(\frac{3x}{2}\right)\cos\left(\frac{x}{2}\right).$$

(b) We have
$$\sin^2(x) + \cos^2(x) = 1.$$

Dividing both sides by $\cos^2(x)$ we get

$$\frac{\sin^2(x)}{\cos^2(x)} + \frac{\cos^2(x)}{\cos^2(x)} = \frac{1}{\cos^2(x)}$$
$$\tan^2(x) + 1 = \sec^2(x)$$

34. Find a simplified formula in terms of $\cos t$ for the expression $\cos 3t$. [Hint: $3t = 2t + t$.]

ANSWER:

$$\cos 3t = \cos(2t + t) \text{ (by hint)}$$
$$= \cos 2t \cos t - \sin 2t \sin t \text{ (using identity)}$$
$$= (2\cos^2 t - 1)\cos t - (2\sin t \cos t)\sin t \text{ (double angle formula)}$$
$$= 2\cos^3 t - \cos t - 2\sin^2 t \cos t$$
$$= 2\cos^3 t - \cos t - 2(1 - \cos^2 t)\cos t$$
$$= 2\cos^3 t - \cos t - 2\cos t + 2\cos^3 t$$
$$= 4\cos^3 t - 3\cos t$$

35. Starting from the difference of angle formulas for sine and cosine, prove the negative angle identities:

(a) $\sin(-t) = -\sin t$ \qquad\qquad **(b)** $\cos(-t) = \cos t$

ANSWER:

(a)

$$\sin(-t) = \sin(0 - t)$$
$$= \sin 0 \cos t - \cos 0 \sin t$$
$$= -\sin t$$

(b)

$$\cos(-t) = \cos(0 - t)$$
$$= \cos 0 \cos t + \sin 0 \sin t$$
$$= \cos t$$

36. Consider the function $f(t) = \left(e^{-0.1t} + 5\right) \sin t$.

(a) What happens to the value of $f(t)$ as t gets very large (that is, as $t \to \infty$)?

(b) From what you know about the behavior of the functions $y = e^{-0.1t}$ and $y = \sin t$, explain why this makes sense. (A sketch might be helpful.)

ANSWER:

(a) From the graph in Fig 7.14 we see that as $t \to \infty$ the graph of $f(t)$ begins to look more and more like a trigonometric function, more specifically it looks like $y = 5 \sin t$.

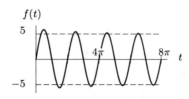

Figure 7.14

(b) We know that

$$f(t) = \left(e^{-0.1t} + 5\right) \sin t = e^{-0.1t} \sin t + 5 \sin t$$

let

$$g(t) = e^{-0.1t} \sin t \quad \text{and} \quad h(t) = 5 \sin t$$

Then we have

$$f(t) = g(t) + h(t).$$

Now we know that as $t \to \infty$ we will have $g(t) \to 0$ since $g(t)$ is an exponential decay function. Thus when t is very large we have

$$f(t) = g(t) + h(t) = \text{small number} + h(t) \approx h(t) = 5 \sin t.$$

37. Two weights (weight 1 and weight 2) are suspended from the ceiling by springs. At time $t = 0$ (t in seconds), the weights are set in motion and begin bobbing up and down. Eventually, however, the oscillation of both weights dies down. The following equations describe the distance of each weight from the ceiling as a function of time:

$$d_1 = 6 + 4\cos(\pi t)e^{-0.2t} \text{ and } d_2 = 5 + 3\cos(2\pi t)e^{-0.5t}.$$

Answer the following question about the weights.

(a) Which weight (number 1 or 2), when at rest, is farthest from the ceiling?

(b) Which weight is closer to the ceiling at time $t = 2$?

(c) Which weight bobs up and down the fastest?

(d) The oscillations of which weight die down the fastest?

(e) At what time are the two weights farthest apart?

ANSWER:

These two graphs should be useful when answering the questions.

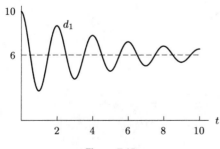

Figure 7.15

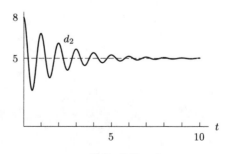

Figure 7.16

(a) Based on the graphs, as the oscillations die down, weight 1 comes to rest at about 6 cm from ceiling and weight 2 at about 5 cm. Thus, at rest, weight 1 is farthest from ceiling.

(b) At $t = 2$, $d_1 \approx 8.68$ and $d_2 \approx 6.10$, So weight 2 is closer.

(c) Weight 1 bobs up and down about once every 2 seconds, and weight 2 bobs up and down about once per second. So weight 2 bobs faster.

(d) Judging from the graphs, the oscillation of weight 2 dies down the fastest.

(e) The distance between weight 1 and weight 2 is greatest when the difference between d_1 and d_2 is greatest. The figure below gives a graph of $y = d_1 - d_2$. As you can see, the two weights are about 4.18 cm apart at $t \approx 0.365$, which is the greatest distance between them. (At $t \approx 0.985$, they are almost this far apart, but not quite.)

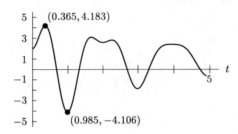

Figure 7.17

38. Table 38 gives $A(t)$, the percentage of the electorate favoring candidate A during the 12 months preceding a presidential election. Time, t, is measured in months and $t = 0$ is a year before election day.

Table 7.3

t (months)	0	3	6	9	12
$A(t)$ percentage favoring candidate A	22	42	62	42	22

(a) Find a trigonometric formula for $A(t)$.

(b) Support for candidate B is given by the formula $B(t) = 31 + 15\sin\left(\frac{\pi}{6}t\right)$.

   (i) At what approximate times are the candidates tied for electoral support? What percentage of the electorate supports each candidate at these times?

   (ii) Sketch a graph of the function $f(t) = A(t) + B(t)$ for $0 \le t \le 12$. What does this function represent about the candidates?

   (iii) Give the approximate maximum and minimum values of $f(t)$ from part (ii). Interpret the meaning of these values in reference to candidates A and B.

   ANSWER:

(a) The points in Table 0.1 have been plotted in Figure 7.18, and a cosine curve has been dotted in.

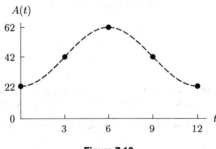

Figure 7.18

Based on the Figure 7.18, the curve has an amplitude of 20, a midline of 42 and a period of 12. This gives

$$A(t) = 42 - 20\cos\left(\frac{\pi}{6}t\right).$$

(b) (i) Setting $A(t) = B(t)$, we can solve the equation

$$42 - 20\cos\left(\frac{\pi}{6}t\right) = 41 + 15\sin\left(\frac{\pi}{6}t\right)$$

graphically. We find two solutions, $t \approx 3.36$ and $t \approx 11.1$. (See Figure 7.19.) The electoral support for each candidate at these times is 45.7% and 24.2% respectively.

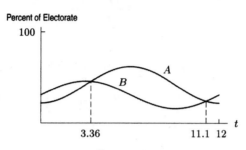

Figure 7.19

(ii) The function $f(t) = A(t) + B(t)$ tells us the total electoral support for both candidates combined.

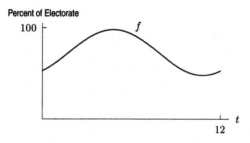

Figure 7.20: $f(t) = A(t) + B(t)$

(iii) Judging from the graph, we have a maximum value for f of about 98% and a minimum of about 48%. This means that at its peak, the combined popularity of the two candidates reaches 98%, while at its low point, the combined popularity reaches 48%.

39. Constructive and destructive interference are defined in Figure 7.21 which is from an engineering textbook.[1] The following four sinusoidal functions (called waves by an engineer) will be used to find examples of destructive and constructive interference:

$$f(x) = \sin^2(x)$$
$$g(x) = \cos^2(x)$$
$$h(x) = 1 + \sin^2(x)$$
$$i(x) = 2\cos^2(x) + \sin^2(x)$$

There are six pairs of the above four functions (e.g. $f(x)$ and $g(x)$, $f(x)$ and $h(x)$, ..., $h(x)$ and $i(x)$.) For each pair, state whether the two waves have destructive interference, constructive interference or neither.

[1]James D. Rancourt, *Optical Thin Films* (New York: McGraw-Hill Publishing Company, 1987).

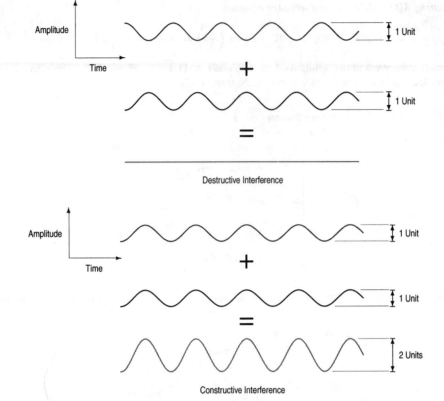

Figure 7.21: Constructive and destructive interference of light waves. The amplitudes of two waves with the same frequency (wavelength) add if the two waves are in phase, or cancel if they have the same amplitudes and are exactly out of phase with one another. In other cases of unequal amplitudes or inexact phasing, the result is a wave with a different amplitude from the original two interfering waves. More than two waves may interfere.

ANSWER:

(a) The first pair, $f(x)$ and $g(x)$, is destructive since

$$\sin^2(x) + \cos^2(x) = 1.$$

(b) $f(x)$ and $h(x)$ are constructive as $\sin^2(x)$ and $1 + \sin^2(x)$ both have the same frequency, and the two are in phase with each other.

(c) $f(x)$ and $i(x)$ are destructive as

$$\sin^2(x) + 2\cos^2(x) + \sin^2(x) = 2\sin^2(x) + 2\cos^2(x)$$
$$= 2\left(\sin^2(x) + \cos^2(x)\right)$$
$$= 2$$

Alternatively,

$$2\cos^2(x) + \sin^2(x) = \cos^2(x) + \left(\cos^2(x) + \sin^2(x)\right)$$
$$= \cos^2(x) + 1$$

so that

$$\sin^2(x) + 2\cos^2(x) + \sin^2(x) = \sin^2(x) + \cos^2(x) + 1 = 2$$

(d) $g(x)$ and $h(x)$ are destructive as

$$\cos^2(x) + 1 + \sin^2(x) = 2$$

(e) $g(x)$ and $i(x)$ are constructive as $g(x) = \cos^2(x)$ and $i(x) = 1 + \cos^2(x)$ have the same frequency and the two are in phase with each other.

(f) $h(x)$ and $i(x)$ are destructive as

$$1 + \sin^2(x) + 2\cos^2(x) + \sin^2(x) = 1 + 2\sin^2(x) + 2\cos^2(x)$$
$$= 1 + 2\left(\sin^2(x) + \cos^2(x)\right)$$
$$= 3$$

Alternatively, if we write $i(x) = 1 + \cos^2(x)$

$$1 + \sin^2(x) + 1 + \cos^2(x) = 3.$$

40. What are the polar and Cartesian equations of the circle in Figure 7.22?

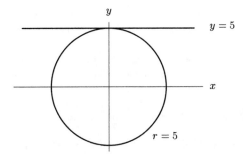

Figure 7.22

ANSWER:
Polar equation: $r = 5$
Cartesian equation: $x^2 + y^2 = 25$

41. (a) What is the Cartesian equation of the circle in Figure 7.23?
(b) Check that all the points satisfying $r = -2\sin\theta$, for $\pi \le \theta \le 2\pi$, lie on this circle.

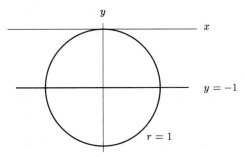

Figure 7.23

ANSWER:
(a) Cartesian equation is unit circle $x^2 + y^2 = 1$ shifted down by 1, giving $x^2 + (y+1)^2 = 1$.
(b) Since θ is restricted to $\pi \le \theta \le 2\pi$, the value of r is positive (as it should be) and the graph is below the x-axis.
If $r = -2\sin\theta$, then

$$x = r\cos\theta = -2\sin\theta\cos\theta$$
$$y = r\sin\theta = -2\sin\theta\sin\theta = -2\sin^2\theta.$$

Substituting into the Cartesian equation gives

$$x^2 + (y+1)^2 = (-2\sin\theta\cos\theta)^2 + (-2\sin^2\theta + 1)^2$$
$$= 4\sin^2\theta\cos^2\theta + 4\sin^4\theta - 4\sin^2\theta + 1$$
$$= 4\sin^2\theta(\cos^2\theta + \sin^2\theta) - 4\sin^2\theta + 1$$
$$= 4\sin^2\theta - 4\sin^2\theta + 1$$
$$= 1.$$

Thus, the points which satisfy $r = -2\sin\theta$ lie on the circle.

42. The origin is at the center of a clock, with the positive x−axis going through 3 and the positive y−axis going through 12. The hour hand is 7 cm long and the minute hand is 9 cm long. What are the Cartesian coordinates and the polar coordinates of the tips of the hour and minute hand, H and M respectively, at the following times?

(a) 2:00 am   **(b)** 6:15 pm

ANSWER:

(a) At 2:00, M is pointing straight up. Thus its rectangular coordinates are (0,9) and its polar coordinates are $(\pi/2, 9)$. H is pointing towards the 2, so its polar coordinates are $(\pi/3, 7)$. Converting this to rectangular coordinates gives $(7\sqrt{3}/2, 7/2)$.

(b) At 6:15, M is pointing towards the 3, so its rectangular coordinates are (9,0) and its polar coordinates are (0,9). H is pointing one fourth of the way between the 6 and the 7. In polar coordinates, the angle between each number is $\pi/6$. Thus, H is at an angle of

$$\frac{3\pi}{4} \text{ (the angle at 6 o}'\text{clock)} - \frac{\frac{\pi}{6}}{4} \text{ (the additional 15 minutes)} = \frac{17\pi}{24}$$

This gives polar coordinates of $(17\pi/24, 7)$ and rectangular coordinates of $(7\cos(17\pi/24), 7\sin(17\pi/24))$.

43. Give inequalities for r and θ which describe the following region in polar coordinates.

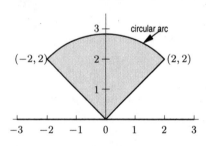

Figure 7.24

ANSWER:
$0 \le r \le 2$ and $\pi/4 \le \theta \le 3\pi/4$

44. Graph the equation $r = n\sin\theta$ for $n = 1, 2, 3, 4$. What is the relationship between the value of n and the shape of the graph?

ANSWER:
As n increases, the period and midline remain the same but the amplitude increases.

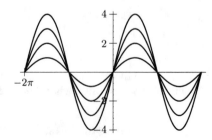

Figure 7.25

45. **(a)** Write $1 - i$ in polar form.

    (b) Find a cube root of $1 - i$. Write your answer in Cartesian form, rounded to two decimal places.

    ANSWER:

(a) $\sqrt{2}e^{7\pi i/4}$

(b) We have

$$
\begin{aligned}
(\sqrt{2}e^{7\pi i/4})^{1/3} &= 2^{1/6}e^{7\pi i/12} \\
&= 2^{1/6}(\cos\frac{7\pi}{12} + i\sin\frac{7\pi}{12}) \\
&= -0.29 + 1.08i.
\end{aligned}
$$

46. The complex number $2 - 3i$ is one root of a quadratic equation with real coefficients. Find one such equation.

    ANSWER:

    One such equation can be obtained by using the product of $x - (2 - 3i)$ and its conjugate. We have

$$
\left(x - (2-3i) \right) \cdot \overline{\left(x - (2-3i) \right)} = x^2 - 4x + 13 = 0
$$

47. Use Euler's formula to derive the identity $\cos\theta = \sin(\theta + \pi/2)$. (Note that if a, b, c, d are real numbers, $a + bi = c + di$ means that $a = c$ and $b = d$.)

    ANSWER:

    Using Euler's formula,

$$
e^{i(\theta+\pi/2)} = \cos(\theta + \pi/2) + i\sin(\theta + \pi/2)
$$

Also, using exponent rules,

$$
\begin{aligned}
e^{i(\theta+\pi/2)} &= e^{i\theta} \cdot e^{i\pi/2} \\
&= (\cos\theta + i\sin\theta)(\cos\pi/2 + i\sin\pi/2) \\
&= (\cos\theta + i\sin\theta)i \\
&= \sin\theta + i\cos\theta.
\end{aligned}
$$

Setting the imaginary parts equal, we get

$$
\cos\theta = \sin(\theta + \pi/2)
$$

Chapter 8 Exam Questions

Exercises

1. Complete the following table for $m(x) = f(g(x))$ and $n(x) = g(f(x))$.

Table 8.1

x	0	1	2	3	4	5
$f(x)$	2	0	5	1	3	4
$g(x)$	1	2	4	0	5	3
$m(x)$						
$n(x)$						

ANSWER:

Table 8.2

x	0	1	2	3	4	5
$f(x)$	2	0	5	1	3	4
$g(x)$	1	2	4	0	5	3
$m(x)$	0	5	3	2	4	1
$n(x)$	4	1	3	2	0	5

2. Find a formulas for $f(g(x))$ and $g(f(x))$ when $f(x) = \dfrac{1}{x}$ and $g(x) = x^2 - 1$.

ANSWER:

$$f(g(x)) = f(x^2 - 1)$$
$$= \frac{1}{x^2 - 1}$$
$$g(f(x)) = g\left(\frac{1}{x}\right)$$
$$= \left(\frac{1}{x}\right)^2 - 1$$
$$= \frac{1}{x^2} - 1$$

3. Find a formulas for $m(n(x))$ and $n(m(x))$ when $m(x) = e^x$ and $n(x) = \dfrac{x^2}{x+1}$.

ANSWER:

$$m(n(x)) = m\left(\frac{x^2}{x+1}\right)$$
$$= e^{\frac{x^2}{x+1}}$$
$$n(m(x)) = n(e^x)$$
$$= \frac{(e^x)^2}{e^x + 1}$$
$$= \frac{e^{2x}}{e^x + 1}$$

4. (a) If $u(v(x)) = \dfrac{3}{1 + \sqrt{x}}$, find possible formulas for $u(x)$ and $v(x)$.

(b) If $f(x) = x^2 + 5x$ and $g(x) = 2 - x$, find a formula for $f(g(x))$.

(c) If $p(q(x)) = \dfrac{6}{2 + x}$ and $q(x) = 1 + x$, find a formula for $p(x)$.

ANSWER:

(a) Since $\dfrac{3}{1 + \sqrt{x}}$ is not a simple function of x, we can break it down into simpler parts. One way to do this is to let

$v(x) = 1 + \sqrt{x}$ and let $u(x) = \dfrac{3}{x}$. That way, when we plug $v(x)$ into $u(x)$, we get the desired result. (There are others.)

(b) We have the formula $f(x) = x^2 + 5x$. To find $f(g(x))$, which is $f(2 - x)$, we plug in $2 - x$ wherever the formula for f has an x. This gives $f(g(x)) = f(2 - x) = (2 - x)^2 + 5(2 - x) = (4 - 4x + x^2) + (10 - 5x) = x^2 - 9x + 14$.

(c) We can solve this by writing the formula for $p(q(x))$ so that the term $1 + x$ appears. We can write $p(q(x)) = \dfrac{6}{2 + x} = \dfrac{6}{1 + (1 + x)}$. Then we replace $1 + x$ with $q(x)$, since we were given the information that $q(x) = 1 + x$.

This gives $p(q(x)) = \dfrac{6}{1 + q(x)}$. Now, to find $p(x)$, replace $q(x)$ with x in that formula. This gives the final result

$$p(x) = \frac{6}{1 + x}.$$

5. Use a graph to decide whether or not the function $y = \left| \dfrac{1}{x} \right|$ is invertible.

ANSWER:

The function is not invertible (it fails the horizontal line test).

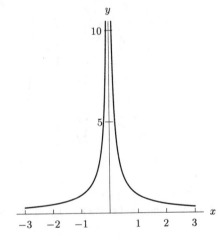

Figure 8.1

6. Check that the functions $f(x) = \dfrac{x^3 + 1}{3}$ and $g(t) = \sqrt[3]{3t - 1}$ are inverses.

ANSWER:

First check $f(g(x))$:

$$f(g(x)) = f(\sqrt[3]{3x - 1})$$
$$= \frac{(\sqrt[3]{3x - 1})^3 + 1}{3}$$
$$= \frac{(3x - 1) + 1}{3}$$
$$= \frac{3x}{3}$$
$$= x.$$

Similarly,

$$g(f(x)) = g\left(\frac{x^3 + 1}{3}\right)$$

$$= \sqrt[3]{3\left(\frac{x^3+1}{3}\right)-1}$$
$$= \sqrt[3]{(x^3+1)-1}$$
$$= \sqrt[3]{x^3}$$
$$= x.$$

7. Find the inverse of $h(x) = e^{\sqrt{x}-1}$.

ANSWER:

Solve $y = e^{\sqrt{x}-1}$ for x :

$$y = e^{\sqrt{x}-1}$$
$$\ln y = \sqrt{x} - 1$$
$$\ln y + 1 = \sqrt{x}$$
$$(\ln y + 1)^2 = x$$

Thus, $h^{-1}(x) = (\ln x + 1)^2$.

8. Find the inverse of $f(x) = \dfrac{3x+1}{x-2}$.

ANSWER:

Solve $y = \dfrac{3x+1}{x-2}$ for x :

$$y = \frac{3x+1}{x-2}$$
$$y(x-2) = 3x+1$$
$$yx - 2y = 3x+1$$
$$yx - 3x = 2y+1$$
$$x(y-3) = 2y+1$$
$$x = \frac{2y+1}{y-3}$$

Thus, $h^{-1}(x) = \dfrac{2x+1}{x-3}$.

In Exercises 9–12, find a simplified formula for the function. Let $f(x) = x+2$, $g(x) = x^3$, and $h(x) = \sqrt{x-1}$.

9. $2f(x) - g(x)$

ANSWER:

$$2f(x) - g(x) = 2(x+2) - x^3$$
$$= -x^3 + 2x + 4$$

10. $f(x)g(x)$

ANSWER:

$$f(x)g(x) = (x+2)x^3$$
$$= x^4 + 2x^3$$

11. $\dfrac{f(x)}{g(h(x))}$
   ANSWER:

$$\frac{f(x)}{g(h(x))} = \frac{x+2}{(\sqrt{x-1})^3}$$
$$= \frac{x+2}{(x-1)^{3/2}}$$

12. $(h(f(x)))^2$
   ANSWER:

$$(h(f(x)))^2 = (\sqrt{(x+2)-1})^2$$
$$= x+1$$

Problems

13. Given the graphs of f and g below, answer the following.

   (a) Evaluate $f(g(3))$
   (b) Evaluate $f^{-1}(0)$
   (c) Estimate $f^{-1}(-2)$
   (d) Solve $g(x) = 1$
   (e) Solve $f(x) = x$

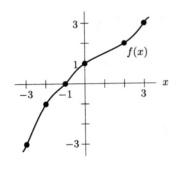
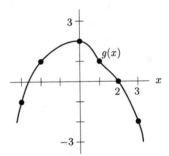

   ANSWER:
   (a) Since $g(3) = -2$, we know $f(g(3)) = f(-2) = -1$.
   (b) $f^{-1}(0)$ is the x-value which gives $f(x) = 0$. Looking at the figure, the answer is $x = -1$. Thus, $f^{-1}(0) = -1$.
   (c) We can approximate which x-value gives $f(x) = -2$ from the graph of $f(x)$. It is about $x = -2.5$, so $f^{-1}(-2) \approx -2.5$.
   (d) From the graph, $g(-2) = 1$ and $g(1) = 1$, so $x = -2$ and $x = 1$ are solutions.
   (e) From the graph, $f(-3) = -3$, $f(2) = 2$, and $f(3) = 3$, so $x = -3$, $x = 2$, and $x = 3$ are solutions.

14. $H(x) = e^{2x-3}$. Find two functions f and g so that $H = f(g(x))$. (Do not choose either $f(x) = x$ or $g(x) = x$.)
   ANSWER:
   We want $H(x) = f(g(x))$ where f and g are two functions of x and $f(x) \neq x$, $g(x) \neq x$. Let's break up $H(x)$ into an expression that resembles a composite function: One choice is $f(x) = e^x$ and $g(x) = 2x - 3$. Another choice is derived from

$$e^{2x-3} = (e^{x-3/2})^2.$$

Thus we can also have $f(x) = x^2$ and $g(x) = e^{x-3/2}$.

15.

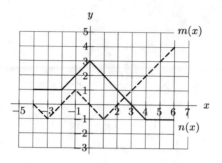

Figure 8.2

The functions $m(x)$ and $n(x)$ are defined by the graph in Figure 8.2. The dotted graph is $m(x)$ and the solid graph is $n(x)$.

(a) Sketch the graph of $y = m(x)$ and the graph of $y = -2m(x + 3) + 1$ on one set of axis.

(b) Sketch the graphs of $y = m(x), y = n(x)$ and of $y = m(x) - n(x)$ on one set of axis.

(c) Evaluate $m(n(3))$ and $n(m(3))$.

(d) Evaluate $m(-1) \cdot n(-1)$ and $m(n(-1))$.

ANSWER:

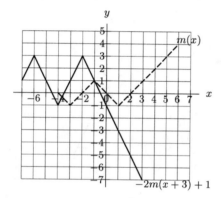

Figure 8.3

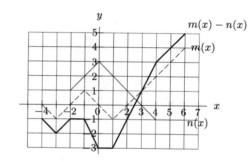

Figure 8.4

(a) See Figure 8.3.

(b) See Figure 8.4.

(c) $n(3) = 0$ so $m(n(3)) = m(0) = 0$.
$m(3) = 1$ so $n(m(3)) = n(1) = -1$.

(d) $m(-1) \cdot n(-1) = (1)(2) = 2$.
$n(-1) = 2$ so $m(n(-1)) = m(2) = 0$.

16. Let $f(x) = 2^x$ and $g(x) = f(f(x))$.

(a) Evaluate the following exactly: $f(0)$, $g(0)$, and $g(g(0))$.

(b) Evaluate the following exactly: $f^{-1}(1)$, $f^{-1}(16)$, and $g^{-1}(16)$. [Note that you don't necessarily need a formula for $f^{-1}(x)$ to answer parts (b) or (c).]

(c) Evaluate the following: $f^{-1}(5)$ and $g^{-1}(5)$. Your answers should either be exact or correct to at least two decimals.

ANSWER:

(a) $f(0) = 2^0 = 1$.
$g(0) = f(f(0))$, but $f(0) = 1$, and so

$$g(0) = f(1) = 2^1 = 2.$$

$g(g(0)) = g(2)$ because $g(0) = 2$. Thus, $g(g(0)) = g(2) = f(f(2)) = f(4)$, because $f(2) = 2^2 = 4$. Therefore,

$$g(g(0)) = f(4) = 2^4 = 16.$$

(b) $f^{-1}(1) = 0$ because $f(0) = 1$.

$f^{-1}(16) = 4$ because, as we saw in part (a), $f(4) = 16$.

In part (a) we saw that $g(g(0)) = 16$, but we also saw that $g(g(0)) = g(2)$. Thus, $g(2) = 16$, and so $g^{-1}(16) = 2$.

(c) If $x = f^{-1}(5)$, then $f(x) = 5$. Thus,

$$2^x = 5$$
$$\ln 2^x = \ln 5$$
$$x \ln 2 = \ln 5$$
$$x = \frac{\ln 5}{\ln 2} \approx 2.322$$

Now, if $x = g^{-1}(5)$ then $g(x) = 5$. Thus, since $g(x) = f(f(x)) = f(2^x) = 2^{2^x}$, we see that

$$2^{2^x} = 5$$
$$\ln 2^{2^x} = \ln 5$$
$$2^x \cdot \ln 2 = \ln 5$$
$$2^x = \frac{\ln 5}{\ln 2}$$
$$\ln 2^x = \ln \left(\frac{\ln 5}{\ln 2} \right)$$
$$x \ln 2 = \ln \left(\frac{\ln 5}{\ln 2} \right)$$
$$x = \frac{\ln \left(\frac{\ln 5}{\ln 2} \right)}{\ln 2} \approx 1.215$$

Note that both of the above solutions can be found (approximately) by using graphs.

17. Figure 8.5 gives the graphs of three different functions, f, which is quadratic, and g and h, which are linear. Sketch possible graphs of the following functions:

$$y = g(f(x)) \qquad \text{and} \qquad y = f(h(x)).$$

Label the axes of each graph accordingly. Note from the graph that $f(b) = a$.

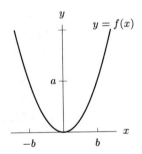 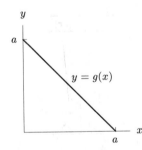 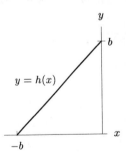

Figure 8.5: Three different functions, one quadratic and two linear

ANSWER:

We see that the slope of g is given by

$$m = \frac{\Delta y}{\Delta x} = \frac{-a}{a} = -1.$$

The y-intercept of g is a, and so a formula for g is $g(x) = a - x$. This means that the equation $y = g(f(x))$ can be written $y = a - f(x)$, or as $y = -f(x) + a$. The graph of this equation, pictured in Figure 8.6, resembles the graph of f flipped over the x-axis and shifted up by a units. Thus, the vertex is at $(0, a)$ and the zeros are at $(-b, 0)$ and $(b, 0)$.

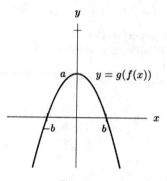

Figure 8.6

As for h, using similar reasoning we see that a formula for h is $h(x) = x + b$. Thus, the equation $y = f(h(x))$ can be written $y = f(x + b)$. The graph of this equation, pictured in Figure 8.7, resembles the graph of f shifted to the left through b units. It has its vertex at $(-b, 0)$ and a y-intercept of $(0, a)$.

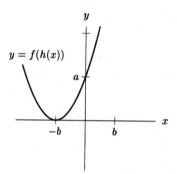

Figure 8.7

18. If $h(x) = f(g(x))$, complete the following tables:

x	$f(x)$
-2	2
-1	0
0	?
1	3
2	-1
5	9
?	6

x	$g(x)$
-1	0
0	?
?	2
2	5
-0.3	1
7	-4
11	-8

x	$h(x)$
-1	-2
1	-1
0	0
2	?
?	3
7	6

ANSWER:

x	$f(x)$
-2	2
-1	0
0	-2
1	3
2	-1
5	9
-4	6

x	$g(x)$
-1	0
0	-1
1	2
2	5
-0.3	1
7	-4
11	-8

x	$h(x)$
-1	-2
1	-1
0	0
2	9
-0.3	3
7	6

19. For each of the following statements, write *true* if the statement is always true, *false* otherwise.

(a) If $y = f(g(x)) = g(f(x))$ then $g^{-1}(f^{-1}(y)) = f^{-1}(g^{-1}(y))$. [Assume that both f and g are invertible functions.]

(b) If f is invertible and $f(x) \to \infty$ as $x \to \infty$ then $f^{-1}(x)$ has a horizontal asymptote.

(c) If $f(x) \cdot g(x)$ is an odd function, then both $f(x)$ and $g(x)$ are odd functions.

(d) If f and g are functions defined for all real values of x, and if $f(x)$ is even, then $h(x) = g(f(x))$ is even.

ANSWER:

(a) Always true. Apply f^{-1} then g^{-1} to both sides of $y = f(g(x))$ to get x. Next apply g^{-1} then f^{-1} to both sides of $y = g(f(x))$ to get x. Thus, $g^{-1}(f^{-1}(y)) = x = f^{-1}(g^{-1}(y))$.

(b) Sometimes true (e.g. $f(x) = \log x$.)

(c) Never true. Let $h(x) = f(x) \cdot g(x)$ and consider what happens when both f and g are odd. $h(-x) = f(-x) \cdot g(-x) = -f(x) \cdot -g(x) = h(x)$, so h is even.

(d) Always true. Consider $h(-x) = g(f(-x)) = g(f(x)) = h(x)$, so h is even.

20. Let f be defined by the graph in Figure 8.8.

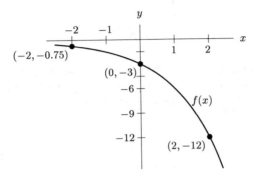

Figure 8.8: The graph of $y = f(x)$

Graph the following equations on separate sets of axes. For each graph, label at least three points, as well as any asymptotes.

(a) $y = -2f(x)$

(b) $y = f(-x) + 2$

(c) $y = f^{-1}(x)$

ANSWER:

(a) This function will resemble an up-side-down version of the graph of $f(x)$, stretched by a factor of two.

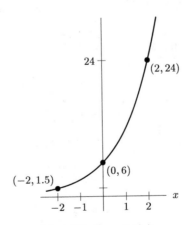

Figure 8.9: $y = -2f(x)$

(b) This graph will resemble the graph of $f(x)$ flipped about the y-axis and up-shifted by two units.

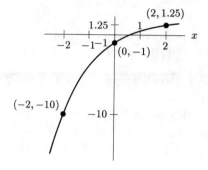

Figure 8.10: $y = f(-x) + 2$

(c) This graph will resemble the graph of $f(x)$ flipped about the line $y = x$.

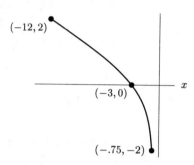

Figure 8.11: $y = f^{-1}(x)$

21. Let $f(x) = \dfrac{1}{1+x}$ and let $g(x) = f(f(x))$.

(a) Find a formula for $f^{-1}(x)$, expressed as one simplified fraction.

(b) Find a formula, in terms of x, for $g^{-1}(x)$, expressed as one simplified fraction.

ANSWER:

(a) Let $y = \dfrac{1}{1+x}$ and express x in terms of y.

$$y = \frac{1}{1+x}$$
$$1 + x = \frac{1}{y}$$
$$x = \frac{1}{y} - 1$$
$$x = \frac{1-y}{y}.$$

Switching x and y,

$$y = \frac{1-x}{x}.$$

So

$$f^{-1}(x) = \frac{1-x}{x}$$

(b)

$$f(f(x)) = g(x) = \frac{1}{1 + \dfrac{1}{1+x}} = \frac{x+1}{x+2}$$

Let $y = \dfrac{x+1}{x+2}$. Then

$$(x+2)y = x+1$$
$$xy + 2y = x + 1$$
$$xy - x = 1 - 2y$$
$$x = \frac{1-2y}{-1+y}.$$

Switch x and y

$$y = \frac{1-2x}{-1+x}$$

So

$$g^{-1}(x) = \frac{1-2x}{-1+x}.$$

22. Given

$$f^{-1}(x) = 200(1.05)^x.$$

(a) Find a formula for $f(x)$.
(b) Solve $f^{-1}(x) = 250$ exactly.
(c) Solve $f(x) = 0$ exactly.

ANSWER:

(a) We are asked to find the inverse of

$$y = 200(1.05)^x.$$

Solving for x:

$$y = 200(1.05)^x$$
$$\ln y = \ln 200 + x \ln 1.05$$
$$x = f(y) = \frac{\ln y - \ln 200}{\ln 1.05}.$$

Thus, writing the formula for f in terms of x, we have

$$f(x) = \frac{\ln x - \ln 200}{\ln 1.05}.$$

(b) We are asked to solve for x such that

$$f^{-1}(x) = 250,$$

in other words, we need to find x such that

$$f(250) = x,$$

which is just evaluating $f(250)$. Thus,

$$f(250) = \frac{\ln 250 - \ln 200}{\ln 1.05} = \frac{\ln(250/200)}{\ln 1.05} = \frac{\ln 1.25}{\ln 1.05}.$$

(c) We are asked the value of x for which

$$f(x) = 0.$$

In other words, we are asked to evaluate (exactly) the value of

$$f^{-1}(0).$$

Substituting 0 into our formula for $f^{-1}(x)$ we get

$$f^{-1}(0) = 200(1.05)^0 = 200(1) = 200.$$

23. Find a formula for $f^{-1}(x)$ if $f(x) = \dfrac{3 + \ln(x-5)}{3 - \ln(x-5)}$.

ANSWER:

Solving for x, we have

$$y = \frac{3 + \ln(x-5)}{3 - \ln(x-5)}$$
$$3y - y\ln(x-5) = 3 + \ln(x-5)$$
$$3y - 3 = y\ln(x-5) + \ln(x-5)$$
$$3y - 3 = (y+1)\ln(x-5)$$
$$\ln(x-5) = \frac{3y-3}{y+1}$$
$$x - 5 = e^{\frac{3y-3}{y+1}}$$
$$x = f^{-1}(y) = e^{\frac{3y-3}{y+1}} + 5.$$

Rewriting f^{-1} in terms of x, we have

$$f^{-1}(x) = e^{\frac{3x-3}{x+1}} + 5.$$

24. Let $f(x) = x^2 + 1$; $g(x) = \frac{1}{x} + 2$ and $h(x) = \sqrt{x - 2}$. Find

(a) $f(x+h)$ (b) $g(f(x))$ (c) $h(f(x))$

(d) $f(h(x))$ (e) $f(3) + g(4)$ (f) $g^{-1}(x)$

ANSWER:

(a) $(x+h)^2 + 1 = x^2 + 2xh + h^2 + 1$

(b) $g(x^2 + 1) = \dfrac{1}{x^2 + 1} + 2 = \dfrac{1 + 2(x^2 + 1)}{x^2 + 1} = \dfrac{2x^2 + 3}{x^2 + 1}$

(c) $h(x^2 + 1) = \sqrt{x^2 + 1 - 2} = \sqrt{x^2 - 1}$

(d) $f(\sqrt{x-2}) = (\sqrt{x-2})^2 + 1 = x - 2 + 1 = x - 1$

(e) $(3^2 + 1) + (\frac{1}{4} + 2) = 10 + 2 + \frac{1}{4} = 12.25$

(f) Let $g(x) = y$ and we have

$$y = \frac{1}{x} + 2$$
$$y - 2 = \frac{1}{x}$$
$$x = g^{-1}(y) = \frac{1}{y - 2}$$
$$g^{-1}(x) = \frac{1}{x - 2}$$

25. For the function g shown below, either sketch a graph of g^{-1} on the same set of axes, or if the function is not *invertible*, give a brief explanation of why it is not invertible.

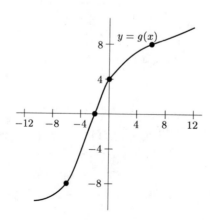

ANSWER:

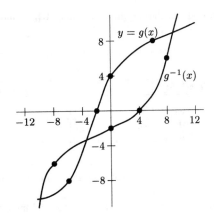

26. The table gives three functions, f, g, and h.

x	f	g	h
-2	2	4	6
-1	4	16	20
0	6	36	42
1	8	64	72
2	10	100	110

(a) Find a possible formula for $f(x)$ in terms of x.
(b) Find a possible formula for $g(x)$ in terms of $f(x)$.
(c) Find a possible formula for $h(x)$ in terms of $f(x)$ and $g(x)$.
(d) Find a possible formula for $h(x)$ in terms of x.

ANSWER:

(a) Looking at the values of $f(x)$ as x goes from -2 to 2 we note that for every unit increment of x, $f(x)$ increases by two units. Thus the formula for $f(x)$ in terms of x will be a linear one with slope 2. We also know that when $x = 0$ we have $f(x) = 6$. Thus the y-intercept of $f(x)$ is 6 and the formula for $f(x)$ in terms of x is

$$f(x) = 2x + 6.$$

(b) Comparing $f(x)$ to $g(x)$ we note that
$$g(x) = (f(x))^2.$$

(c) Looking at the sum of $f(x)$ and $g(x)$ we note that

$$h(x) = f(x) + g(x).$$

(d) Since $f(x) = 2x + 6$ and $g(x) = (f(x))^2$, we get

$$g(x) = (2x + 6)^2.$$

From part (c) we know that
$$h(x) = f(x) + g(x).$$

Thus using what we know about $f(x)$ and $g(x)$ we get

$$h(x) = 2x + 6 + (2x + 6)^2.$$

Simplifying we get
$$h(x) = 2x + 6 + 4x^2 + 24x + 36 = 4x^2 + 26x + 42.$$

27. Table 8.3 describes the growth of frequent-flier programs at major US airlines. In these programs, airlines award free tickets to program members on the basis of the number of miles that they fly. The table gives values for $P(t)$, the number of individual members enrolled in frequent-flier programs at major US airlines, in millions, and $M(t)$, the cumulative number of miles earned by all members, in billions. For both functions, t gives the number of years since 1981.

Table 8.3

t, years since 1981	0	2	4	6	8	10	12
$P(t)$, individual members (millions)	2	5	7	12	20	28	30
$M(t)$, cumulative miles earned (billions)	100	250	400	700	800	1000	1700

(a) Define $A(t)$ to be the average number of accumulated miles per member (in thousands). Give a formula for $A(t)$ in terms of $P(t)$ and $M(t)$.

(b) During the period from 1981 to 1993, when does $A(t)$ reach its minimum value?

ANSWER:

(a) We have

$$A(t) = \text{average number of accumulated miles per member (in thousands)}$$
$$= \frac{\text{total number of accumulated miles (in billions)}}{\text{total number of members (in millions)}}$$
$$= \frac{M(t)}{P(t)}.$$

Notice that 1 billion/1 million $= 1$ thousand, and so the units are correct.

(b) Table 8.4 gives values of $A(t)$.

Table 8.4

t (years since 1981)	0	2	4	6	8	10	12
$A(t) = M(t)/P(t)$ (average number of miles per member (in thousands))	50	50	57.1	58.3	40	35.7	56.7

From the table, we see that $A(t)$ appears to reach a minimum value in year 10, or 1991.

28.

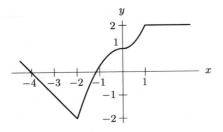

Figure 8.12

The function $h(x)$ is shown in Figure 8.12. Match each of the functions (a)–(d) with the graph that best matches it.

(a) $y = 1/h(x)$
(b) $y = -h(x)$
(c) $y = |h(x)|$
(d) $y = h(-x)$

I.

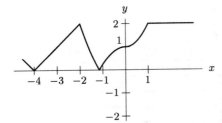

Figure 8.13

II.

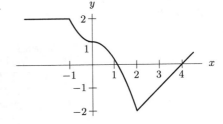

Figure 8.14

III.

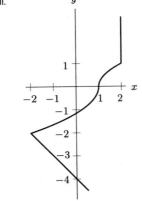

Figure 8.15

IV.

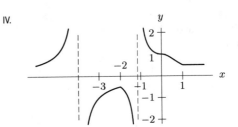

Figure 8.16

V.

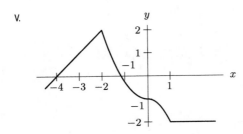

Figure 8.17

ANSWER:

(a) IV

(b) V

(c) I

(d) II

29. Table 8.5 gives values of the functions f, g, and h, three functions defined only for $x = 0, 1, \ldots, 5$. Based on the table, answer the following questions.

(a) Construct a table of values for $u(x) = 2f(x) + f(g(x))$.

(b) State the relationship between $h(x)$ and $f(x)$.

Table 8.5

x	0	1	2	3	4	5
f	3	2	4	0	1	5
g	5	1	4	3	0	2
h	3	4	1	0	2	5

ANSWER:

(a)

Table 8.6

x	$2f(x)$	$f(g(x))$	$u(x) = 2f(x) + f(g(x))$
0	6	5	11
1	4	2	6
2	8	1	9
3	0	0	0
4	2	3	5
5	10	4	14

(b) Notice that :

$f(0) = 3$ and $h(3) = 0$
$f(1) = 2$ and $h(2) = 1$
$f(2) = 4$ and $h(4) = 2$
$f(3) = 0$ and $h(0) = 3$
$f(4) = 1$ and $h(1) = 4$
$f(5) = 5$ and $h(5) = 5$
Thus,

$$h(x) = f^{-1}(x).$$

Chapter 9 Exam Questions

Exercises

1. Are the following functions power functions? If so, write them in the form $y = kx^p$.

    (a) $y - 4 = (x + 2)(x - 2)$ **(b)** $y = (x - 3)^2$

    ANSWER:

  (a) Yes.

$$y - 4 = (x + 2)(x - 2) = x^2 - 4$$
$$y = x^2$$

  (b) No.

2. Find a power function through the points $(1, 4)$ and $(4, 7)$.

    ANSWER:

    We want a function of the form $y = kx^p$. Thus $4 = k \cdot 1^p = k$. Substituting this in with the other point, we get

$$7 = 4(4)^p = 4^{p+1}$$
$$\ln 7 = \ln 4^{p+1} = (p + 1) \ln 4$$
$$p + 1 = \frac{\ln 7}{\ln 4}$$
$$p = \frac{\ln 7}{\ln 4} - 1 = 0.404$$

  Thus, $y = 4x^{0.404}$.

3. Suppose y is directly proportional to x. If $y = 4$ when $x = 2$, find the constant of proportionality and write the formula for y as a function of x. Use your formula to find x when y is 6. Repeat the problem for when y is inversely proportional to x.

    ANSWER:

    When y is directly proportional to x, we know $y = kx$. Thus $4 = 2k$, so $k = 2$ and the formula is $y = 2x$. Thus $x = 3$ when $y = 6$.

    When y is inversely proportional to x, we know $y = k/x$. Thus $4 = 2/k$, so $k = 8$ and the formula is $y = 8/x$. Thus $x = 4/3$ when $y = 6$.

4. Find a possible formula for the power function.

Table 9.1

x	-1	0	1	2
$f(x)$	-2	0	2	1

    ANSWER:

    We know $y = kx^p$. When $x = 1$, we get $2 = k(1)^p = k$. Thus, $y = 2x^p$. Plugging in the value when $x = 2$, we get

$$1 = 2(2)^p = 2^{p+1}$$

  Since $1 = 2^0$, we have $p = -1$. Thus, $y = 2x^{-1}$.

5. Are the following functions polynomials? If so, of what degree?

    (a) $y = 4 + 3x$ **(b)** $y = x^2 + e^x + 1$ **(c)** $y = 3.1x^9 - 4.7$

    ANSWER:

  (a) Yes, degree 1
  (b) No, e^x is not a part of a polynomial
  (c) Yes, degree 9

6. Describe the *end behavior* of the function $f(x) = -2x^4 - 3x^3 + 3x - 5$ by completing the following statements:

   (a) As $x \to \infty$, $f(x) \to \ldots$?
   (b) As $x \to -\infty$, $f(x) \to \ldots$?

   ANSWER:

   (a) $-\infty$
   (b) $-\infty$

7. Find the zeros of the function $y = x^3 - 2x^2 - 8x$.

   ANSWER:

$$y = x^3 - 2x^2 - 8x$$
$$= x(x - 4)(x + 2)$$

   Thus, the zeros are at $x = 0, 4$, and -2.

8. Use a graphing calculator or computer to graph $f(x) = x^4 + 2x^3 - 7x^2 - 8x + 12$, and use this graph to determine the factored form of $f(x)$.

   ANSWER:

   From the graph, we see that $f(x)$ has zeros at $x = -3, -2, 1$, and 2. Thus,

$$f(x) = (x + 3)(x + 2)(x - 1)(x - 2).$$

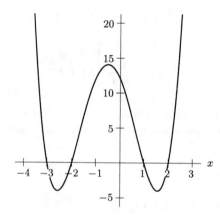

Figure 9.1

9. Find the zeros of the function $y = x^3 - 2x^2 - 8x$.

   ANSWER:

$$y = x^3 - 2x^2 - 8x$$
$$= x(x - 4)(x + 2)$$

   Thus, the zeros are at $x = 0, 4$, and -2.

10. Without a calculator, graph $f(x) = \frac{1}{3}(x^2 - 9)(x + 2)$. Label all the x−intercepts and y−intercepts.

    ANSWER:

    First note that $f(0) = \frac{1}{3}(-9)(2) = -6$, so the y−intercept will be at $y = -6$. Next, factor $f(x)$ to find its zeros:

$$f(x) = \frac{1}{3}(x + 3)(x - 3)(x + 2).$$

    Thus, the zeros are at $x = -3, -2$, and 3.

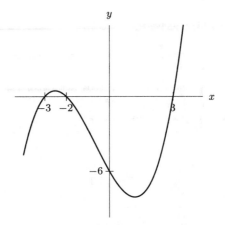

Figure 9.2

Are the functions in Exercises 11–13 rational functions? If so, write them in the form $p(x)/q(x)$, the ratio of polynomials.

11. $f(x) = \dfrac{3}{x^2} + \dfrac{x+1}{2}$
ANSWER:

$$\frac{3}{x^2} + \frac{x+1}{2} = \frac{3(2)}{2x^2} + \frac{x^2(x+1)}{2x^2}$$
$$= \frac{x^3 + x^2 + 6}{2x^2}$$

Thus, $f(x)$ is rational.

12. $f(x) = \dfrac{x^2 + 2^x}{x}$
ANSWER:
Not rational.

13. $f(x) = \dfrac{1}{x+1} - \dfrac{x}{x-1}$
ANSWER:

$$\frac{1}{x+1} - \frac{x}{x-1} = \frac{x-1}{(x+1)(x-1)} - \frac{x(x+1)}{(x+1)(x-1)}$$
$$= \frac{-x^2 - 1}{x^2 - 1}$$

14. Find the x–intercepts, y–intercepts, and horizontal and vertical intercepts (if any) of

$$f(x) = \frac{x^2 - 1}{x^2 + 4x}$$

ANSWER:
First factor the numerator and denominator:

$$f(x) = \frac{x^2 - 1}{x^2 + 4x} = \frac{(x+1)(x-1)}{x(x+4)}.$$

Setting $x = 0$ gives an undefined fraction, so there are no y–intercepts. Setting $f(x) = 0$ gives x–intercepts of $x = \pm 1$. To find horizontal asymptotes, look at the long-run behavior:

$$f(x) = \frac{x^2 - 1}{x^2 + 4x} \approx \frac{x^2}{x^2} = 1.$$

Thus, the horizontal asymptote is at $y = 1$. To find vertical asymptotes, look at the zeros of the denominator. There are vertical asymptotes at $x = 0$ and $x = 4$.

15. Graph $y = \dfrac{x}{(x^3 - 3x + 2)}$; labeling all asymptotes.

ANSWER:

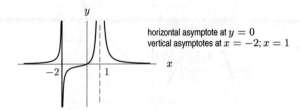

horizontal asymptote at $y = 0$
vertical asymptotes at $x = -2; x = 1$

Figure 9.3

16. Consider the function

$$f(x) = \frac{-20(x + \frac{1}{2})(x - 3)}{(x + 7)(x - 15)}$$

(a) What are the x-intercepts of f?
(b) What is the y-intercept of f?
(c) What are the vertical asymptotes of f?
(d) What is the horizontial asymptote of f?
(e) Sketch a graph of f, showing clearly all the features you calculated in (a)–(d).

ANSWER:

(a) $x = -1/2, 3$
(b) $y = \dfrac{-20(1/2)(-3)}{7(-15)} = -\dfrac{2}{7}$
(c) $x = -7, x = 15$
(d) $y = -20$
(e)

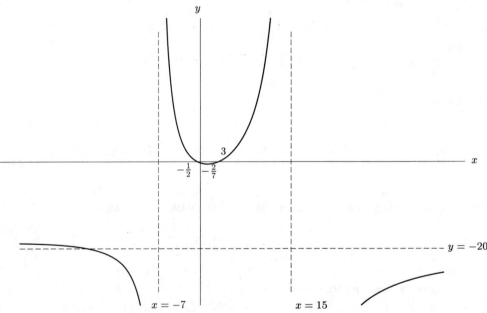

17. Let $f(x) = \dfrac{3}{x+1}$.

(a) Complete the following table to determine the behavior of $f(x)$ as x approaches -1 from the right and from the left.

Table 9.2

x	-2	-1.1	-1.01	-1	-0.99	-0.9	0
$f(x)$							

(b) Complete the following tables to determine the behavior of $f(x)$ as x takes very large and very small values.

Table 9.3

x	5	10	100	1000
$f(x)$				

Table 9.4

x	-5	-10	-100	-1000
$f(x)$				

(c) Using the information from the tables, graph $f(x)$ without a calculator.

ANSWER:

(a) From the left, $f(x)$ is approaching $-\infty$. From the right, $f(x)$ is approaching ∞.

Table 9.5

x	-2	-1.1	-1.01	-1	-0.99	-0.9	0
$f(x)$	-3	-30	-300	undefined	300	30	3

(b) As x takes on very large positive values, $f(x)$ approaches 0 from the positive side.

Table 9.6

x	5	10	100	1000
$f(x)$	0.5	0.2727	0.0297	0.0030

As x takes on very large negative values, $f(x)$ approaches 0 from the negative side.

Table 9.7

x	-5	-10	-100	-1000
$f(x)$	-0.75	-0.3333	-0.0303	-0.0030

(c)

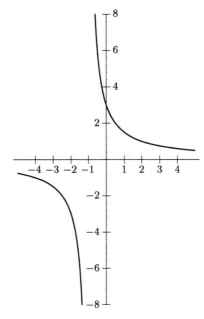

Figure 9.4

18. Can the following functions be written in the form of an exponential function or a power function?

(a) $f(x) = x^3(x-1)^4$ (b) $g(x) = 2 \cdot 3^{2-x}$ (c) $h(x) = \dfrac{5}{3x^2}$

ANSWER:

(a) No.

(b) Yes. $f(x) = 2 \cdot 3^2 \cdot 3^{-x} = 18 \left(\dfrac{1}{3}\right)^x$.

(c) Yes. $h(x) = \dfrac{5}{3}x^{-2}$.

19. Let $f(x) = x^{1/3}$ and $g(x) = \ln x$. Complete the following table of values and describe the long-run behaviors of f and g as $x \to \infty$.

Table 9.8

x	2	4	6	8	10	100	1000
$f(x)$							
$g(x)$							

ANSWER:

The power function $f(x) = x^{1/3}$ eventually dominates the logarithmic function $g(x) = \ln x$.

Table 9.9

x	2	4	6	8	10	100	1000
$f(x)$	1.26	1.59	1.82	2	2.15	4.64	10
$g(x)$	0.69	1.39	1.79	2.08	2.30	4.60	6.91

20. Which function dominates as $x \to \infty$?

$$y = 1.1^x \qquad y = (10x)^{1.1}$$

ANSWER:

We know that the exponential function $y = 1.1^x$ will eventually dominate the power function $y = (10x)^{1.1} = 12.589x^{1.1}$.

21. Find a formula for the power function $f(x)$.

Table 9.10

x	2	3	4	5
$f(x)$	12.865	47.089	118.228	241.453

ANSWER:

Use power function regression on a calculator, or solve algebraically:

We know $f(x) = ax^b$ for some constants a and b. Using the first two points, $12.865 = a(2)^b$ and $47.089 = a(3)^b$.

Thus,

$$\frac{12.865}{2^b} = \frac{47.089}{3^b}$$

$$12.865(3)^b = 47.089(2)^b$$

$$\left(\frac{3}{2}\right)^b = \frac{47.089}{12.865} = 3.6599$$

$$b \ln\left(\frac{3}{2}\right) = \ln 3.6599$$

$$b = 3.2$$

To solve for a, again use any point:

$$12.865 = a(2)^{3.2}$$

$$a = \frac{12.865}{2^{3.2}}$$

$$= 1.4$$

Thus, $f(x) = 1.4x^{3.2}$.

22. Find a formula for the exponential function $g(x)$.

Table 9.11

x	2	3	4	5
$g(x)$	3.4461	3.6874	3.9455	4.2217

ANSWER:

Use exponential function regression on a calculator, or solve algebraically:

We know $g(x) = ab^x$ for some constants a and b. Using the first two points, $3.4461 = a(b)^2$ and $3.6874 = a(b)^3$.

Thus,

$$\frac{3.4461}{b^2} = \frac{3.6874}{b^3}$$
$$3.4661b^3 = 3.6874b^2$$
$$3.4661b = 3.6874$$
$$b = 1.07$$

To solve for a, again use any point:

$$3.4461 = a(1.07)^2$$
$$a = 3.01$$

Thus, $g(x) = 3.01(1.07)^x$.

23. Find a formula for the polynomial function of degree 2.

Table 9.12

x	1	2	3	4
$h(x)$	0	3	10	21

ANSWER:

Use regression on a calculator or solve algebraically:

We know $h(x) = ax^2 + bx + c$ for some constants a, b and c. Using the first three points,

$$0 = a + b + c$$
$$3 = 4a + 2b + c$$
$$10 = 9a + 3b + c$$

Subtracting the first equation from the second and the second from the third gives

$$3 = 3a + b$$
$$7 = 5a + b$$

Thus, $b = 3 - 3a = 7 - 5a$, so $a = 2$. Substituting back in,

$$b = 3 - 3(2) = -3$$

and

$$0 = 2 - 3 + c, \text{ so } c = 1.$$

Thus, $h(x) = 2x^2 - 3x + 1$.

Problems

24. According to a recent advertisement, Burger King's all-beef hamburger patties have 75% more beef than McDonald's all-beef hamburger patties. If both chains serve circular hamburger patties of the same thickness, then the diameter of the Burger King patties, d_B, will be directly proportional to the diameter of the MacDonald's patties, d_M. Express d_B as a function of d_M.

ANSWER:

Since hamburger patties are cylindrical in shape the volume of a McDonald patty is:

$$V_m = \pi r_m{}^2 h = \pi \left(\frac{d_m}{2}\right)^2 h$$

$$\text{now } V_b = 1.75 V_m = 1.75\pi \left(\frac{d_m}{2}\right)^2 h$$

$$\text{but } V_b = \pi \left(\frac{d_b}{2}\right)^2 h$$

$$\text{therefore } \pi \left(\frac{d_b}{2}\right)^2 h = 1.75\pi \left(\frac{d_m}{2}\right)^2 h$$

$$\text{or } \frac{d_b{}^2}{4} = 1.75 \frac{d_m{}^2}{4}$$

$$\text{or } d_b = 1.75^{\frac{1}{2}} d_m$$

25. The volume occupied by a fixed quantity of gas such as oxygen is inversely proportional to its pressure, provided that its temperature is held constant. Suppose that a quantity of oxygen occupies a 100-liter volume at a pressure of 15 atmospheres. If the temperature of the oxygen does not change, what volume will it occupy if its pressure rises to 19 atmospheres?

ANSWER:

Let V be the volume and P the pressure of the oxygen. Since V is inversely proportional to P, we know that $V = \frac{k}{P}$ from some constant k. Since 100 liters of the oxygen is at a pressure of 15 atmospheres, this gives

$$100 = \frac{k}{15}$$

and so $k = 1500$. If the pressure rises to 19 atmospheres, we have

$$V = \frac{k}{19} = \frac{1500}{19} \approx 78.9 \text{ liters.}$$

26. When temperature is held constant, the pressure P and volume V of a quantity of gas are inversely proportional (Boyle's Law). Figure 9.5 shows the relationship between P and V for a particular gas.

(a) Fill in Table 9.13.

(b) Write a formula for V in terms of P.

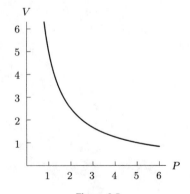

Figure 9.5

Table 9.13

P (atm)	V(cm^3)
	5
10	
	20

ANSWER:

(a) Since P is inversely proportional to V we can write

$$P = \frac{k}{V} \quad \text{and} \quad V = \frac{k}{P}$$

where k is a constant. Thus $PV = k$. From the graph, when $P = 1$, $V = 5$ and when $P = 2.5$, $V = 2$, so $k = 5$. Using this information, we can now find the missing values in the table.

For $V = 5\text{cm}^3$, $P = \frac{k}{5} = \frac{5}{5} = 1$atm.
For $P = 10$atm, $V = \frac{k}{10} = \frac{5}{10} = 0.5\text{cm}^3$.
For $V = 20\text{cm}^3$, $P = \frac{k}{20} = \frac{5}{20} = 0.25$atm.
So we have

Table 9.14

P (atm)	$V(\text{cm}^3)$
1	5
10	0.5
0.25	20

(b) Using the value of k found in (a),

$$P = \frac{5}{V}$$

or

$$PV = 5.$$

27. Poiseuille's law says that the rate at which blood is flowing through a blood vessel of radius R is proportional to R^4. For medical reasons, we want to know how a reduction in the radius of the blood vessel affects the blood flow.

(a) If the radius of the the blood vessel decreases by 10%, by what percentage does the blood flow decrease?
(b) If the radius of the blood vessel decreases by 50%, by what percentage does the blood flow decrease?

ANSWER:

(a) If the rate at which blood is flowing is B, we know

$$B = kR^4.$$

If R is replaced by $0.9R$, we have

$$B = k(0.9R)^4 = (0.9)^4 kR^4 = 0.6561kR^4.$$

Thus, B is multiplied by 0.6561, or decreased by $0.34 \approx 34\%$.

(b) If R is replaced by $0.5R$ we have

$$B = k(0.5R)^4 = (0.5)^4 kR^4 = 0.0625kR^4.$$

Thus, B is multiplied by 0.0625, or decreased by $0.94 \approx 94\%$.

28. Match a graph to each of the following statements:

    (a) p is proportional to the square root of q. Graph _____

    (b) p is inversely proportional to the square of q. Graph _____

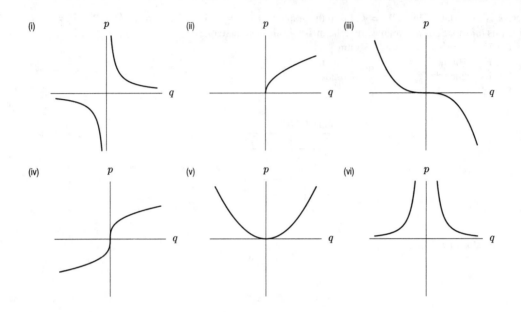

       ANSWER:

(a) (ii)

(b) (vi)

29. Figure 9.6 gives the graphs of two power functions, f and g. Find a possible formula for $g(x)$.

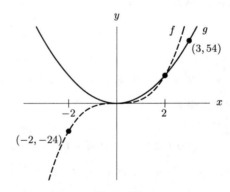

Figure 9.6

       ANSWER:

    The point $(-2, -24)$ is on the graph of f. By symmetry, the point $(2, 24)$ is also on this graph. Since f and g intersect at $x = 2$, we know that g contains the points $(2, 24)$ and $(3, 54)$. This allows us to find a formula for g, namely $g(x) = 6x^2$.

30. Figure 9.7 gives graphs of $f(x) = ax^p$ and $g(x) = bx^q$.

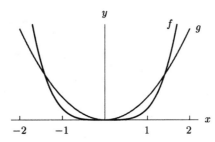

Figure 9.7: Graphs of $f(x) = ax^p$ and
$g(x) = bx^q$

(a) Which is larger, p or q? Explain your answer.
(b) Which is larger, a or b? Explain your answer.
(c) If $f(g(1)) = (ab)^p$, what is a?

ANSWER:

(a) Note that as $x \to \infty$, f climbs faster than g. Thus, $p > q$.
(b) Note that $f(1) = a \cdot 1^p = a$ and $g(1) = b \cdot 1^p = b$. Since at $x = 1$ the graph of g lies above the graph of f, we see that $b > a$.
(c)

$$f(g(1)) = f(b \cdot 1^q) = f(b) = a \cdot b^p$$
$$\text{so} \quad a \cdot b^p = (ab)^p$$
$$a \cdot b^p = a^p \cdot b^p$$
$$a^{p-1} = 1.$$

This equation is true if $a = 1$ or $p = 1$. However, p can not be 1 because $f(x)$ is not linear; thus, $a = 1$.

31. Suppose that a, b, c, and d are integers, and that a is positive and even, b is positive and odd, c is negative and even, and d is negative and odd. Figure 9.8 gives the graphs of four power functions. For each of the equations that follow, state which of the graphs it could describe, if any.

(i) 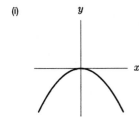 (ii) 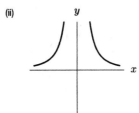 (iii) 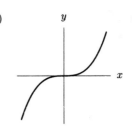 (iv)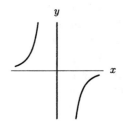

Figure 9.8: The graphs of four power functions

(a) $y = cx^a$
(c) $y = bx^a$
(e) $y = cx^d$

(b) $y = ax^c$
(d) $y = bx^b$
(f) $y = dx^c$

ANSWER:

(a) Graph (i)
(b) Graph (ii)
(c) No match
(d) Graph (iii)
(e) Graph (iv)
(f) No match

32. Figure 9.9 gives the graph of a function, $f(x)$, whose formula has the form $f(x) = k \cdot x^p$.

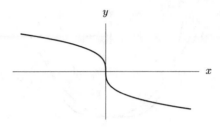

Figure 9.9

Circle all statements that must be true:

(a) $k > 1$ (b) $0 < k < 1$

(c) $k < 0$ (d) k must be an integer

(e) $p > 1$ (f) $0 < p < 1$

(g) $p < 0$ (h) p must be an integer

ANSWER:

The true statements are (c) and (f).

33. Figure 9.10 gives the graphs of four power functions. For each of the equations that follow, state which of the graphs it could possibly describe. Note that some of the equations may not have a matching graph.

(i) 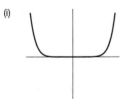   (ii) 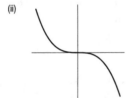   (iii) 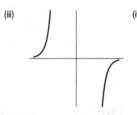   (iv)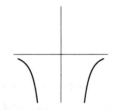

Figure 9.10: The graphs of four different power functions

(a) $y = -\dfrac{1}{100x^4}$ (b) $y = -10x^3$

(c) $y = 1000x^8$ (d) $y = 50x^{-2}$

(e) $y = -0.001x^{-7}$ (f) $y = \pi x^{13}$

ANSWER:

(a) Graph (iv)

(b) Graph (ii)

(c) Graph (i)

(d) No match

(e) Graph (iii)

(f) No match

34. Give a possible formula for a power function, $f(x)$, which satisfies all the following conditions:

- $f(-x) = f(x)$
- $\lim_{x \to \infty} f(x) = 0$
- $f(2) = -\frac{3}{4}$

ANSWER:

$f(-x) = f(x)$ implies that f is an even function. Since $\lim_{x \to \infty} f(x) = 0$, we can deduce that f contains a negative exponent of x. We can therefore try $f(x) = k \cdot x^{-2}$. Then for $f(2) = -\frac{3}{4}$, $k = -3$ so one answer that works is

$$f(x) = -3x^{-2}$$

35. Suppose $f(x) = a \cdot x^b$ and $g(x) = c \cdot x^d$ have the graphs in Figure 9.11. Circle whether each of the following statements is True, False, or Could be True or False.

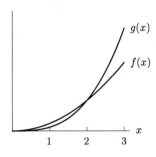

Figure 9.11

$a < b$	True	False	Could be True or False.
$a < c$	True	False	Could be True or False.
$b < d$	True	False	Could be True or False.
$c < d$	True	False	Could be True or False.

ANSWER:

From the graphs, we see that all the constants a, b, c, d, must be positive. Since $g(x)$ is above $f(x)$ for large x, we know $b < d$ must always be true. Since $g(x)$ is below $f(x)$ for $x = 1$, we have $c < a$. So $a < c$ is not true. We cannot say anything about $a < b$ and $c < d$, so they are marked Could be True or False.

$a < b$ Could be True or False

$a < c$ False

$b < d$ True

$c < d$ Could be True or False

36. Suppose each of the polynomials, f, graphed in parts (a)–(f) has leading term ax^n, that is,

$$f(x) = ax^n + \text{ terms of lower degree.}$$

For each function, circle the sign of a, circle whether n is odd or even, and write the minimum possible value of n.

(a)

a is   positive   negative
n is   odd   even
smallest possible n is _____

(b)

a is   positive   negative
n is   odd   even
smallest possible n is _____

(c)

a is   positive   negative
n is   odd   even
smallest possible n is _____

(d)

a is   positive   negative
n is   odd   even
smallest possible n is _____

(e)

a is   positive   negative
n is   odd   even
smallest possible n is _____

(f)

a is   positive   negative
n is   odd   even
smallest possible n is _____

ANSWER:

(a) a is positive
n is odd
$n \geq 3$

(b) a is negative
n is odd
$n \geq 3$

(c) a is positive

n is even

$n \geq 4$

(d) a is positive

n is odd

$n \geq 5$

(e) a is negative

n is even

$n \geq 4$

(f) a is positive

n is odd

$n \geq 3$

37. (a) Use a graphing calculator to find all real zeroes for $f(x) = 0.1x^3 + 1.5x^2 - 0.4x - 6$.

(b) Explain how you know you have found all of the zeroes of the function of part (a).

ANSWER:

(a) zero(s):-15,-2,2

(b) The function is a cubic polynomial. It can have at most 3 real zeroes.

38. The graphs of $y_1 = 0.25x^4$ and $y_2 = 0.25(x^4 + x^3 - 23x^2 + 3x + 90)$ are shown in the figure below as viewed on the window $[-10, 10]$ by $[-25, 25]$.

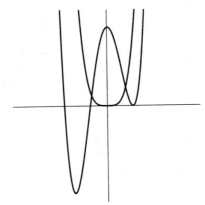

(a) What are the zeroes of

(i) y_1?

(ii) y_2?

(b) Give the dimensions of a viewing window which shows the graphs of y_1 and y_2 as indistinguishable from one another.

ANSWER:

(a) (i) 0

(ii) $-5, -2, 3, 3$

(b) $[-50, 50]$ by $[-100000, 100000]$

39. If $f(x) = x^2 - 11$ and $g(x) = \dfrac{x^4}{2} - 15$, find all x for which $f(x) > g(x)$ both graphically and algebraically.

ANSWER:

To solve the problem algebraically, first set $f(x) = g(x)$:

$$x^2 - 11 = \frac{x^4}{2} - 15$$

$$x^2 = \frac{x^4}{2} - 4$$

$$2x^2 = x^4 - 8$$

$$0 = x^4 - 2x^2 - 8$$

$$= (x^2 - 4)(x^2 + 2)$$

$$= (x - 2)(x + 2)(x^2 + 2)$$

Thus, the functions are equal at $x = -2, 2$. Next, note that at $x = 0$, $f(x) = -11$, and $g(x) = -15$. Thus on the interval $-2 < x < 2$ we have $f(x) > g(x)$. We see the same thing from the graphs of the functions:

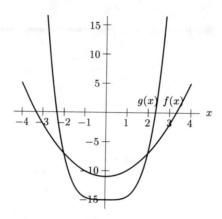

Figure 9.12

40. The function $f(x) = \cos x$ can be approximated by the function $g(x) = 1 - \dfrac{x^2}{2} + \dfrac{x^4}{24}$. Make a table of values to show on what interval this is a good estimate. How could you use this approximation to find $\cos x$ for values not in this range?

ANSWER:

Table 9.15 shows the values of $f(x)$ and $\cos x$ for $\pi < x < \pi$. We can see that $f(x)$ is a very good approximation for $-\frac{\pi}{3} < x < \frac{\pi}{3}$ and a fairly good approximation for $-\frac{\pi}{2} < x < \frac{\pi}{2}$. To approximate other values of $\cos x$, use the fact that the cosine function is periodic and the identity $\cos (x - \pi) = -\cos x$ (which can be derived from the sum-of-angle formulas).

Table 9.15

x	0	$\pm\frac{\pi}{6}$	$\pm\frac{\pi}{3}$	$\pm\frac{\pi}{2}$	$\pm\frac{2\pi}{3}$	$\pm\frac{5\pi}{6}$	$\pm\pi$
$\cos x$	1	0.87	0.5	0	-0.5	-0.87	-1
$f(x)$	1	0.87	0.50	0.02	-0.39	-0.47	0.12

41. One of the following tables of data comes from a linear function, one from an exponential function, and one from a polynomial. Identify which is which and find a formula for each.

(a)

x	$f(x)$
10	112
15	98
20	84
25	70
30	56

(b)

x	$g(x)$
-2	16
-1	24
0	36
1	54
2	81

(c)

x	$h(x)$
-3	-3
-2	0
-1	1
0	0
1	-3

ANSWER:

There are many ways to decide which function is of which type. One way is described here. The data in table (a) shows a constant decrease of 14 units for values of $f(x)$ for each increase of 5 units in values for x. Thus f must be the linear function. The slope, $\frac{\Delta f}{\Delta x}$ would be $m = \frac{-14}{5}$, and the function will be of the form $f(x) = b + mx$. Thus,

$$f(x) = b - \frac{14}{5}x.$$

Use any point from the table to solve for b, e.g., $f(10) = 112$ gives

$$f(x) = b - \frac{14}{5}(10),$$

so

$$b = 112 + 28,$$
$$b = 140.$$

Thus, a formula for the linear function is

$$f(x) = 140 - \frac{14}{5}x.$$

The exponential function will be of the form $y = a \cdot b^x$, and y cannot be zero (unless $a = 0$, in which case the function is always zero). Therefore, we can exclude h from the possibilities for the exponential function. That leaves g. Since $g(0) = 36$, we know $a = 36$. We can use any other point to solve for b. A convenient point would be $g(1) = 54$. Then,

$$54 = 36 \cdot b,$$

$$b = \frac{54}{36} = \frac{3}{2}.$$

A formula for the exponential function is

$$g(x) = 36\left(\frac{3}{2}\right)^x.$$

Table (c) is left for the polynomial function. Note that if we had started with the data from table (c), we could have excluded both the linear and exponential models from the fact that $h(-3) = h(1) = 3$ or from $h(-2) = h(0) = 0$. The points from table (c) are plotted in Figure 9.13 with a curve dashed in.

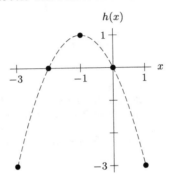

Figure 9.13: $h(x)$ is a polynomial

The graph looks similar to a parabola, which we know to be a function of degree 2, so let's assume the formula looks like $h(x) = ax^2 + bx + c$. To solve for a, b, and c, we'll need three results from the table. First, take $h(0)$; our formula tells us that $h(0) = a(0^2) + b(0) + c = c$. Since $h(0) = 0$, we get $c = 0$. Thus, there is no constant term in the polynomial and we can write it as $h(x) = ax^2 + bx$. Next, consider $h(1)$. The formula gives $h(1) = a(1^2) + b(1) = a + b$; the table gives $h(1) = -3$. Thus, $a + b = -3$. Using $h(-1) = 1$, we find $h(-1) = a(-1)^2 + b(-1) = a - b$. Thus, $a - b = 1$. If we add this equation to the result that $a + b = -3$, we get $2a + b - b = -2$, which means that $2a = -2$, so $a = -1$. Substituting back into $a + b = -3$ gives $-1 + b = -3$, or $b = -3 + 1 = -2$. Using these solutions, the formula $h(x) = ax^2 + b$ becomes

$$h(x) = -x^2 - 2x.$$

This could have been derived using the vertex form or the factored form, since the vertex and zeros are known.

42. Approximate all zeros of the polynomial $f(x) = x^7 + 5x^2 - 4$.

ANSWER:

Since the degree of $f(x)$ is 7, it can have at most 7 zeros. Graphing the function will give an idea where to look for them.

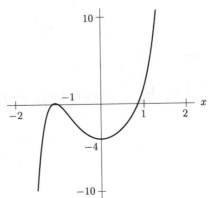

The graph appears to have 3 zeros—two negative zeros very close to each other and one positive zero. From the figure, it looks as though one of the zeros is at -1. Substituting $x = -1$ into the formula for $f(x)$ gives $f(-1) = (-1)^7 + 5(-1)^2 - 4 = -1 + 5 - 4 = 0$, which confirms that $x = -1$ is a zero. A computer or graphing calculator can find approximate values for the other two zeros at $x \approx 0.856$ and $x \approx -1.139$.

43. Find a possible formula for the function graphed below:

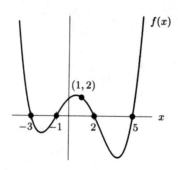

Figure 9.14

ANSWER:

From the graph, we can see that the function, f, has zeroes at $-3, -1, 2$, and 5. Since we can see no other zeroes, the function can possibly be a fourth degree polynomial. Then, the function would have the form:

$$f(x) = k(x + 3)(x + 1)(x - 2)(x - 5).$$

Substituting the point $(1, 2)$, we can determine k.

$$2 = k(4)(2)(-1)(-4)$$
$$k = \frac{1}{16}$$

Thus, $f(x) = \frac{1}{16}(x + 3)(x + 1)(x - 2)(x - 5)$.

44. Let $f(x) = (x - 2)^2(x + 4)$. Sketch the following graphs on separate sets of axes, labeling all axis-intercepts as well as the approximate coordinates of all 'bumps'.

(a) $y = f(x), y = f(-x), y = -f(x), y = -f(-x)$
(b) $y = f(x + 4), y = f(x - 5)$
(c) $y = 2f(x), y = -0.5f(x), y = f(2x), y = f(-x/2)$

ANSWER:

(a)

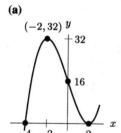

Figure 9.15: $y = f(x) = (x - 2)^2(x + 4)$

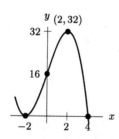

Figure 9.16: $y = f(-x)$

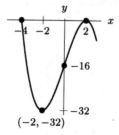

Figure 9.17: $y = -f(x)$

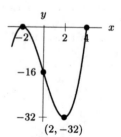

Figure 9.18: $y = -f(-x)$

(b)

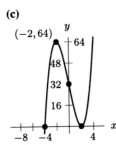

Figure 9.19: $y = f(x+4)$

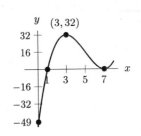

Figure 9.20: $y = f(x-5)$

(c)

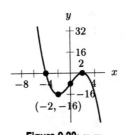

Figure 9.21: $y = 2f(x)$

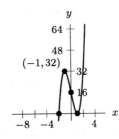

Figure 9.22: $y = -0.5f(x)$

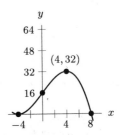

Figure 9.23: $y = f(2x)$

Figure 9.24: $y = f(-\frac{x}{2})$

45. Which of the following could represent a complete graph of $f(x) = ax - x^3$ where a is a constant? Explain your reasoning.

(A)

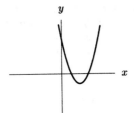

(B)

(C)

(D)

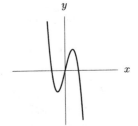

Figure 9.25

ANSWER:

(D).

Graph (A) is an even-degree polynomial, and (C) must be at least a 5th degree polynomial because it has 5 zeros. Graph (B) has a positive leading coefficient. Graph (D) shows the proper behavior for a "flipped" 3rd degree polynomial, with a negative coefficient of x^3.

46. Find a polynomial with integer coefficients having $\dfrac{7 + \sqrt{3}}{5}$ as a zero.

ANSWER:

Since solving for a zero would give

$$x = \frac{7 + \sqrt{3}}{5}$$

we work backward to get a polynomial set to zero

$$5x = 7 + \sqrt{3}$$
$$5x - 7 = \sqrt{3}$$
$$(5x - 7)^2 = 3$$
$$25x^2 - 70x + 49 = 3$$
$$25x^2 - 70x + 46 = 0.$$

47. The volume of pollutants (in millions of cubic feet) in a certain reservoir is given by

$$P(t) = 350 + 30t,$$

where t is in years. The volume of the reservoir (including water and pollutants) is gradually increasing and is given by the formula

$$R(t) = 12000 + 120t.$$

Let $C(t)$ be the fraction of the reservoir's volume that consists of pollutants.

(a) Write an expression for $C(t)$ in terms of t.
(b) In year $t = 0$, what percentage of the reservoir's total volume consists of pollutants?
(c) If these trends were to continue for many many years, about what percentage of the reservoir's total volume would eventually consist of pollutants?

ANSWER:

(a) $C(t) = \dfrac{350 + 30t}{12000 + 120t}$

(b) We have $P(0) = 350$ and $R(0) = 12000$, and so initially the percent of the volume that consists of pollutants is

$$C(0) = \frac{350}{12000} \approx 2.9\%$$

(c) The fraction of the reservoir's volume that consists of pollutants is given by

$$C(t) = \frac{P(t)}{R(t)} = \frac{350 + 30t}{12000 + 120t}.$$

Over time, this fraction would approach

$$\frac{30t}{120t} \approx 25\%$$

and so after many years about 25% of the reservoir's volume would consist of pollutants.

48. A 12 kg sample of a certain alloy (mixture of metals) contains 3 kg of tin and 9 kg of copper. A chemist decides to study the properties of the alloy as its percentage of tin is varied. Suppose x represents the quantity of tin, in kg, the chemist adds to the sample. Let $f(x)$ represent the fraction of the mixture's mass composed of tin — that is, the ratio of the tin's mass to the mixture's total mass. A negative value of x represents a quantity of tin removed from the original 12 kg sample.

(a) Find a formula for f in terms of x.
(b) What is the domain of f?
(c) What is the horizontal asymptote of $y = f(x)$? Explain the physical significance of this asymptote.
(d) Evaluate and interpret the expression $f(0.5)$.
(e) Evaluate and interpret the expression $f^{-1}(0.5)$.

ANSWER:

(a) We know that the total mass of the mixture is the sum of the masses of the tin and the copper. We know that there will always be 9 kg of copper in the mixture and that the mass of tin in the mixture is $3 + x$. Thus the proportion of tin to the total mass is

$$f(x) = \frac{3 + x}{12 + x}.$$

(b) Since there cannot be a negative mass of tin in the alloy, we must have

$$x \geq -3.$$

On the other hand there is no maximum amount of tin that can be in the alloy. Thus, the domain of $f(x)$ is

$$x \geq -3.$$

(c) Looking at the graph of $f(x)$ in Figure 9.26, we see that there is a horizontal asymptote at $y = 1$. This tells us that as we add more and more tin to the mixture, the proportion of tin in the mixture gets closer and closer to one.

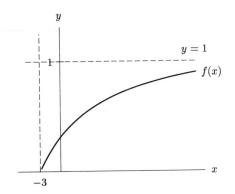

Figure 9.26

(d) We know that

$$f(x) = \frac{3+x}{12+x}.$$

Thus

$$f(0.5) = \frac{3.5}{12.5} = 0.28 \text{(kg tin)/(kg mixture)}.$$

This tells us that after adding 0.5 kilograms of tin to the sample, tin will make up 28% of the mixture.

(e) We are asked to solve for x such that

$$0.5 = \frac{3+x}{12+x}.$$

Solving we get

$$0.5 = \frac{3+x}{12+x}$$
$$6 + 0.5x = 3 + x$$
$$0.5x = 3$$
$$x = 6$$

Thus

$$f^{-1}(0.5) = 6.$$

This tells us that after adding 6 kilograms of tin to the sample, the tin will make up exactly 50% of the mixture.

49. For each function below, complete the statements to describe the end behavior of the function.

(a) $f(x) = -3x^3 + 5x^4 - 2x + \pi$
(b) $g(x) = -6x^5 + 4x^3 - 3$
(c) $h(x) = \dfrac{3x+6}{x+1}$
(d) $j(x) = e^x - 1$

ANSWER:

(a) As $x \to -\infty, f(x) \to \infty$; as $x \to +\infty, f(x) \to \infty$.
(b) As $x \to -\infty, g(x) \to +\infty$; as $x \to +\infty, g(x) \to -\infty$.
(c) As $x \to -\infty, h(x) \to 3$; as $x \to +\infty, h(x) \to 3$.
(d) As $x \to -\infty, j(x) \to -1$; as $x \to +\infty, j(x) \to \infty$.

50. Find possible formulas for the functions in Figure 9.27.

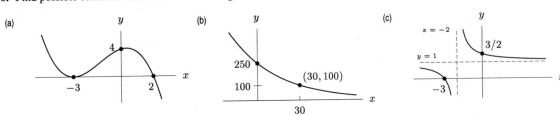

Figure 9.27

ANSWER:

(a) The graph in Figure 9.27 (a) appears to be a third degree polynomial, so let

$$y = k(x+3)^2(x-2),$$

and use $(0, 4)$ to find $k = -\frac{2}{9}$.

(b) The graph in Figure 9.27 (b) looks like exponential decay. Let

$$y = 250 \cdot b^x$$
$$100 = 250 \cdot b^{30}$$
$$b = \left(\frac{100}{250}\right)^{1/30} \approx 0.9699,$$

so

$$y = 250(0.9699)^x.$$

(c) We can view the graph in Figure 9.27 (c) as the graph of $y = \frac{1}{x}$ shifted left 2 units and up one unit. Thus, $y = \frac{1}{x+2} + 1$ or equivalently,

$$y = \frac{x+3}{x+2}.$$

51. The graphs of several different rational functions are described below. Find a possible formula for each of these functions.

(a) The graph of $y = f(x)$ has exactly one vertical asymptote, at $x = -1$, as well as a horizontal asymptote at $y = 1$. The graph of f crosses the y-axis at $y = 3$ and crosses the x-axis exactly once, at $x = -3$.

(b) The graph of $y = g(x)$ has exactly two vertical asymptotes: one at $x = -2$ and one at $x = 3$. It also has a horizontal asymptote at $y = 0$. Furthermore, the graph of g crosses the x-axis exactly once, at $x = 5$.

(c) The graph of $y = h(x)$ has exactly two vertical asymptotes: one at $x = -2$ and one at $x = 3$. It also has a horizontal asymptote at $y = 1$. Furthermore, the graph of h crosses the x-axis exactly once, at $x = 5$.

ANSWER:

(a) The horizontal asymptote of $y = 1$ means the numerator and denominator have the same degree and their leading coefficients match. The zero at $x = -3$ places $(x+3)$ in the numerator, and the vertical asymptote at $x = -1$ places $(x+1)$ in the denominator. So the simplest function satisfying our conditions is

$$f(x) = \frac{x+3}{x+1}.$$

(b) The horizontal asymptote at $y = 0$ means the degree of the denominator is greater that the degree of the numerator. The vertical asymptotes imply the denominator has roots at $x = -2$ and $x = 3$. The zero at $x = 5$ corresponds to the zeros of the numerator. This leads us to

$$g(x) = \frac{(x-5)}{(x+2)(x-3)}.$$

(c) The horizontal asymptote at $y = 1$ means both the degree of the numerator and denominator and their leading coefficients match. The vertical asymptotes correspond to zeros of the denominator at $x = -2$ and $x = 3$. The zero of h at 5 corresponds to the zeros of the numerator. Thus, we expect a factor of $(x-5)$ in the numerator and factors of $(x+2)$ and $(x-3)$ in the denominator. In order to make the degrees of numerator and denominator match *and* make sure $h(x)$ *crosses* the x-axis at $x = 5$, we must square the numerator. So

$$h(x) = \frac{(x-5)^2}{(x+2)(x-3)}$$

52. Find a possible formula for the following function:

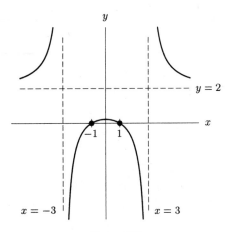

Figure 9.28

ANSWER:

We know that this function has vertical asymptotes at $x = -3$ and $x = 3$. Thus a possible formula for the graph is

$$y = \frac{k}{(x-3)(x+3)} + A = \frac{k}{x^2 - 9} + A.$$

Since the function $y = \frac{k}{x^2-9}$ has a horizontal asymptote at $y = 0$ and the one in Fig 9.28 has one at $y = 2$ we know that the formula for the graph in the figure is a 2-unit vertically shifted version of the function $y = \frac{k}{x^2-9}$. Thus

$$A = 2$$

and the formula for the function in the graph is

$$y = \frac{k}{x^2 - 9} + 2.$$

We know that the function goes through the point $(1, 0)$ so

$$0 = \frac{k}{1^2 - 9} + 2 = 2 - \frac{k}{8}.$$

Solving for k we get

$$k = 16.$$

Thus the formula for the function whose graph is in the figure is

$$y = \frac{16}{x^2 - 9} + 2.$$

53. Graph $f(x) = \dfrac{-2(x+A)(x-C)}{(x+B)(x-D)}$ given that $A > B > C > D > 0$. Label all intercepts and asymptotes.

ANSWER:

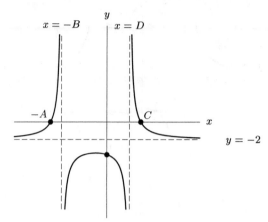

Figure 9.29

y-intercept: $f(0) = \dfrac{-2(A)(-C)}{(B)(-D)} = \dfrac{-2A \cdot C}{B \cdot D} < 0$ since $\dfrac{A \cdot C}{B \cdot D} > 0$.

54. Find possible formulas for each of the following graphs, assuming each of them is the graph of a rational function.

ANSWER:

(a) $f(x) = \dfrac{1}{1 + x^2}$

(b) $g(x) = \dfrac{x}{1 + x^2}$

(c) $h(x) = \dfrac{x^2}{1 + x^2}$

55. Each of the following graphs is a graph of the function of the form

$$f(x) = Mx + \frac{N}{x}, \qquad \text{where } M \text{ and } N \text{ are constants.}$$

For each graph, decide the signs of M and N.

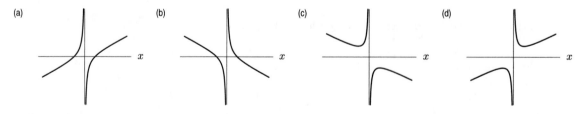

ANSWER:

(a) As $x \to \infty$, $f(x) \to \infty$, so we have $M > 0$.
As $x \to 0^+$, $f(x) \to -\infty$, so $N < 0$.
(b) As $x \to \infty$, $f(x) \to -\infty$, so we have $M < 0$.
As $x \to 0^+$, $f(x) \to \infty$, so $N > 0$.
(c) As $x \to \infty$, $f(x) \to -\infty$, so we have $M < 0$.
As $x \to 0^+$, $f(x) \to -\infty$, so $N < 0$.
(d) As $x \to \infty$, $f(x) \to \infty$, so we have $M > 0$.
As $x \to 0^+$, $f(x) \to \infty$, so $N > 0$.

56. The infant mortality, I, in a country is related to the country's GNP (gross national product), g. Some authors (Weld and Helms, 1997) have argued that the relationship is of the form

$$I = I_0 + \frac{k}{g+a},$$

where I_0, k, and a are positive constants and $g \geq 0$.

(a) Plot a graph of I against g with $I_0 = 2$, $k = 6$, and $a = 3$. Label intercepts and asymptotes.
(b) Does the graph of I against g (for any I_0, k, a) have a horizontal asymptote? If so, what is it?
(c) Does the graph of I against g (for any I_0, k, a) have a vertical asymptote? If so, what is it?
(d) Draw a graph of I against g for any I_0, k, and a. Label intercepts and asymptotes.

ANSWER:

(a)

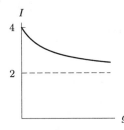

Figure 9.30

(b) Yes; $I = I_0$
(c) No, not for $g \geq 0$. (There is one with $g = -a$.)
(d)

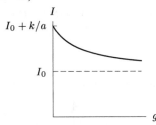

Figure 9.31

57. Tables 9.16, 9.17, and 9.18 give values for 3 functions, f, g, and h, one of which is a power function, one linear, and one exponential. Find possible formulas for these three functions.

Table 9.16

x	$f(x)$
0.11	171
0.31	177
0.41	180
0.71	189
0.81	192

Table 9.17

x	$g(x)$
30	80
35	120
40	180
45	270
50	405

Table 9.18

x	$h(x)$
-2	5/8
-1	5
0	undefined
1	-5
2	$-5/8$

ANSWER:

(a) Looking at the points $(0.11, 171)$, $(0.31, 177)$ and $(0.41, 180)$ in Table 9.16 we see that this is a linear function with slope

$$m = \frac{177 - 171}{0.31 - 0.11} = \frac{180 - 177}{0.41 - 0.31} = 30.$$

Thus it will be of the form

$$f(x) = 30x + b.$$

Substituting the point $(0.11, 171)$ and solving for b we get

$$171 = 30(0.11) + b$$
$$= 3.3 + b.$$

Thus

$$b = 167.7$$

and the function for the values in the table is

$$f(x) = 30x + 167.7.$$

(b) Looking at the points $(30, 80)$, $(35, 120)$ and $(40, 180)$ in Table 9.17 we see that this is an exponential function which increases by a factor of

$$\frac{120}{80} = \frac{180}{120} = 1.5$$

every 5 unit interval. Thus the function will be of the form

$$g(x) = A(1.5)^{x/5}.$$

Substituting the point $(30, 80)$ we get that

$$80 = A(1.5)^{30/5}$$
$$= A(1.5)^6$$
$$\approx A(11.4)$$

Thus

$$A \approx 7.02$$

and the formula for $g(x)$ is

$$g(x) = 7.02(1.5)^{x/5}.$$

This could also be expressed in the form ab^x or ae^{kx}.

(c) This must be the power function. Since it is undefined at some point the power must be negative, so $h(x) = Ax^{-k}$. We also know that it is an odd function since $h(-x) = -h(x)$ Thus we can assume that the function will be of the form

$$h(x) = \frac{A}{x^k},$$

where k is an odd integer. Substituting the point $(-1, 5)$ we get

$$5 = \frac{A}{(-1)^k} = -A \text{ (since } k \text{ is odd)}$$

Thus, $A = -5$. Substituting the point $(2, -\frac{5}{8})$ we get

$$-\frac{5}{8} = -\frac{5}{2^k}$$

or

$$8 = 2^k$$

which gives us

$$k = 3.$$

Thus a possible formula for the function is

$$h(x) = -\frac{5}{x^3}.$$

58. Answer the following questions *yes* or *no*. Give a counter example if your answer is *no*.

(a) Suppose f is a power function and that $f(2) = \dfrac{1}{2}$. Does this mean that $f(x) = \dfrac{1}{x}$?

(b) Does the sum of two functions with horizontal asymptotes also have a horizontal asymptote?

(c) Is it possible that both graphs shown in Figure 9.32 represent exponentially growing populations?

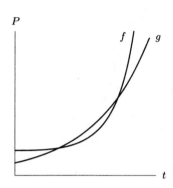

Figure 9.32

ANSWER:

(a) No. For example, $f(x) = \frac{1}{8}x^2$.

(b) Yes.

(c) Yes.

59. Suppose $a, b, c,$ and d are integers, and that a is negative and odd, b is positive and even, c is positive and odd, and d is positive and odd. For each of the six functions (a) – (f) choose one of the graphs I – XII.

(a) $y = -e^x$ corresponds to graph _____

(b) $y = \dfrac{-1}{e^x}$ corresponds to graph _____

(c) $y = -ax^b$ corresponds to graph _____

(d) $y = \dfrac{-c}{x^d}$ corresponds to graph _____

(e) $y = \left| \dfrac{c}{x^d} \right|$ corresponds to graph _____

(f) $y = cx^d$ corresponds to graph _____

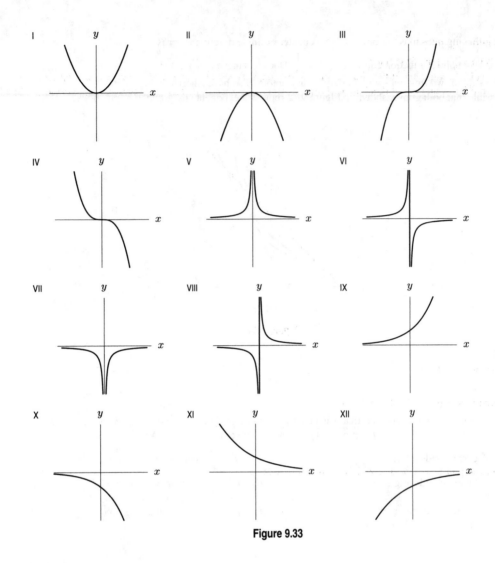

Figure 9.33

ANSWER:

(a) X
(b) XII
(c) I
(d) VI
(e) V
(f) III

60. What is the long-run behavior of the following functions?

(a) $y = \dfrac{3^x - 1}{2^x + 7}$

(b) $y = \dfrac{x^3 + 10}{e^x + 4}$

ANSWER:

(a) For very large positive values of x, $y \approx \dfrac{3^x}{2^x} = \left(\dfrac{3}{2}\right)^x$, which approaches ∞. For very large negative values of x, the terms 3^x and 2^x approach zero. Thus y approaches $-\dfrac{1}{7}$.

(b) For very large positive values of x, $y \approx \dfrac{x^3}{e^x}$. The numerator is power function and the denominator is an exponential function, so the denominator will dominate. Thus $y \to 0$. For very large negative values of x, the numerator approaches $-\infty$ and the denominator approaches 0. Since the denominator again dominates, $y \to -\infty$.

61. Given the following graph

   (a) What is the equation of the line?

   (b) What would the equation of this line be if the horizontal axis was labeled x and the vertical axis labeled $\ln y$?

   (c) What would the equation of this line be if the horizontal axis was labeled $\ln x$ and the vertical axis labeled $\ln y$?

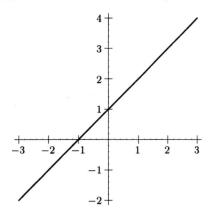

Figure 9.34

     ANSWER:

(a) $y = x + 1$

(b) Replace the y in $y = -2x + 1$ with $\ln y$:

$$\ln y = -2x + 1$$
$$y = e^{-2x+1}$$

(c) Replace the y in $y = -2x + 1$ with $\ln y$ and the x with $\ln x$:

$$\ln y = -2\ln x + 1$$
$$e^{\ln y} = e^{-2\ln x + 1}$$
$$= e \cdot \left(e^{\ln x}\right)^{-2}$$
$$y = ex^{-2}$$

62. The following table represents the amount of monthly data handled by MichNet, the University of Michigan's gateway to the Internet, sampled at six month intervals for the years $1993 - 1995$. (The data is from the *Michigan Daily*, March 20, 1996, p. 4.)

Table 9.19

Date	June '93	Dec. '93	June '94	Dec. '94	June '95	Dec. '95
Data packets (billions)	1.8	2.2	4.1	5.0	7.8	10.5

    In what follows, use d (for data) to represent the number of data packets handled monthly, in billions, and let t represent the number of years since December, 1992. (So June '93 is represented by $t = 0.5$.) *Do not use x and y to represent your variables.* When doing the following parts of this exercise, enter data into your calculator *very* carefully, since one data entry error could result in the loss of many points. It may be helpful as you go along to notice that you will evntually be asked which of the two models you are about to find is better.

   (a) Find an equation that gives the best exponential model for this data, with your final answer being in one of the two forms $d = Ae^{Bt}$ or $d = A \cdot B^t$, with the actual numerical values of A and B filled in. *In this final answer, include as much accuracy for A and B as you can get.*

   (b) Find an equation that gives the best power function model for the data, with your final answer being in the form $d = At^B$ with the actual numerical values of A and B filled in. *In this final answer, include as much accuracy for A and B as you can get.*

(c) Which model do you think is better? Give your reasons; the more convincing, the better your score.

(d) According to the *Daily* article, the data handling capacity of the MichNet system is about 40 billion data packets per month. According to the better model you found above, what is the value of t when the network will become overloaded? What year does this represent?

ANSWER:

(a)

$$\ln d = 0.73219389t + 0.18568868 \quad \text{corr. coeff.} = 0.9919$$

$$e^{\ln d} = e^{0.73219389t + 0.18568868} = e^{0.18568868} e^{0.73219389t}$$

$$d = 1.204047358 e^{0.73219389t} \text{ or } d = 1.204047358(2.079638104)^t$$

(b)

$$\ln d = 0.99975185 \ln t + 1.0637334 \quad \text{corr. coeff.} = 0.9595$$

$$e^{\ln d} = e^{0.99975185 \ln t + 1.0637334} = e^{1.0637334} \left(e^{\ln t}\right)^{0.99975185}$$

$$d = 2.897167107 t^{0.99975185}$$

(c) Though both linearized data sets have correlation coefficients near 1, that of the exponential model is *much* nearer to 1, so that may be the better model.

(d) Using the base e model $1.204047358 e^{0.73219389t}$, we set the exponential function to be equal to 40, and solving for t we get

$$40 = 1.204047358 e^{0.73219389t}$$

$$t = \frac{\ln 40 - \ln 1.204047358}{0.73219389} \approx 4.78$$

Thus, the network will overload during the year $1993 + 4.78 = 1997$

63. Table 9.20 contains the names of the nine known planets that orbit about the sun, along with their distance d from the sun in millions of miles and the period P of the revolution about the sun in standard earth years, that is, the time it takes the planet to go about the sun.

(a) Of the three possible models "linear", "exponential", and "power function", find the one that best fits the data, expressing your answer in the form $P = f(d)$.

Table 9.20

Planet	Distance	Period
Mercury	36.0	0.241
Venus	67.0	0.615
Earth	93.0	1.00
Mars	141.5	1.88
Jupiter	483	11.9
Saturn	886	29.5
Uranus	1782	84
Neptune	2793	165
Pluto	3670	248

(b) Suppose a tenth panet were discovered at a distance of 5 billion (i.e. 5000 million) miles from the sun. Based on your best model found in part (a), predict its period.

ANSWER:

(a) Linear model computation, using the least squares program gives:

$$P = 0.06578634d - 12.50408$$

with a correlation coefficient of 0.9889.

Exponential model computation gives:

$$\ln P = 0.00171667d + 0.28331315$$

with a correlation coefficient of 0.8934.

Power model computation gives:

$$\ln P = 1.4999215 \ln d - 6.796078$$

with a correlation coefficient of 0.99999968.

Comparing correlation coefficients shows that the power function fits best.

Then, to find the formula,

$$e^{\ln P} = e^{1.4999215 \ln d - 6.796078}$$
$$P = e^{-6.796078} \left(e^{\ln d}\right)^{1.4999215}$$
$$P = 0.001118152 d^{1.4999215}$$

(b)

$$P = 0.001118152(5000^{1.4999215}) \approx 395 \text{years}$$

64. Table 9.21 gives population data for two different countries. Find a curve of best fit for this data.

Table 9.21 *Population data for two countries*

year	population 1	population 2
0	100	200
10	170	227
20	320	826
30	513	1707
40	980	3252
50	1357	4552
60	2658	5968
70	3939	8435
80	5118	11819
90	8498	14340
100	13796	16131

ANSWER:

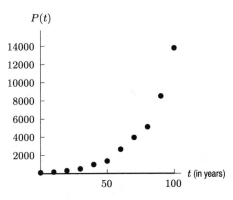

Figure 9.35: Population 1

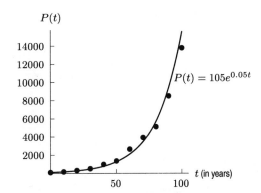

Figure 9.36: Best-fit curve is exponential

In Figure 9.35 we have the data for population 1 plotted on a graph. The best-fit curve in this case would seem to be an exponential curve. In this case, the graph of the function shown in Figure 9.36 is $P(t) = 105e^{0.05t}$. This was obtained "by eye" so your results may differ slightly.

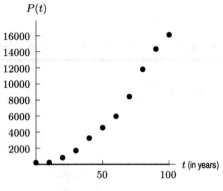

Figure 9.37: Population 2

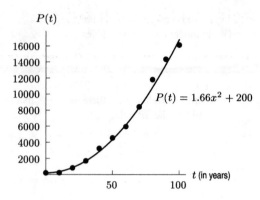

Figure 9.38: Best-fit curve is a power function

In Figure 9.37 we have the data for population 2 plotted on a graph. The best-fit curve, as shown in Figure 9.38, is a power function, whose formula, obtained again "by eye," is given by $P(t) = 1.66x^2 + 200$. Once again, your answer may vary slightly.

Chapter 10 Exam Questions

Exercises

1. Are the following quantities vectors or scalars?

   (a) The velocity of a satellite.

   (b) The distance between the satellite and the earth.

   (c) The gravitational force between the satellite and the earth.

   ANSWER:

   (a) vector

   (b) scalar

   (c) vector

2. Given the displacement vectors $\vec{u}$ and $\vec{v}$ in Figure 10.1, draw the following vectors:

   (a) $\vec{u} + \vec{v}$ (b) $2\vec{u}$ (c) $2\vec{u} - \vec{v}$

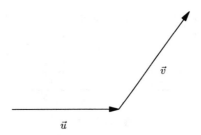

Figure 10.1

ANSWER:

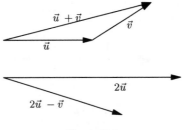
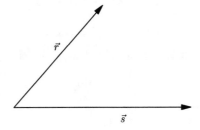

Figure 10.2

3. Given the displacement vectors $\vec{r}$ and $\vec{s}$ in Figure 10.3, draw the following vectors:

   (a) $\vec{r} + \vec{s}$ (b) $-\vec{r}$ (c) $-\vec{r} + 2\vec{s}$

Figure 10.3

ANSWER:

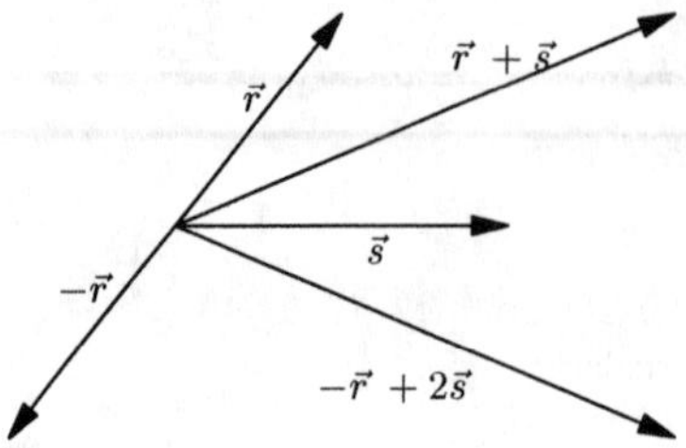

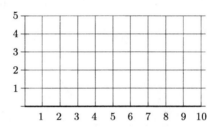

Figure 10.4

4. Perform the computation $(\vec{i} - \vec{j}) - 2(2\vec{i} - \vec{j})$.

ANSWER:
$-3\vec{i} + \vec{j}$

5. On the graph in Figure 10.5, draw the vector $\vec{v} = 2\vec{i} + \vec{j}$ twice, once with its tail at the origin and once with its tail at the point (6,3).

Figure 10.5

ANSWER:

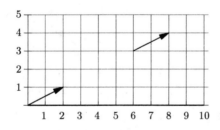

Figure 10.6

6. Resolve the vector starting at the point $P = (1, 1)$ and ending at the point $Q = (3, -2)$ into components.

ANSWER:
$2\vec{i} - 3\vec{j}$

7. Find the length of the vector
$$\vec{v} = -2.1\vec{i} + 3.2\vec{j} - 4.6\vec{k}.$$

ANSWER:
The length is given by $\sqrt{(-2.1)^2 + (3.2)^2 + (-4.6)^2} = 5.984$.

8. Use $\vec{R} = (2, 3, 4, 5, 6)$ and $\vec{S} = (0, 1, 3, 4, 7)$ to find the following vectors.

(a) $\vec{A} = \vec{R} - \vec{S}$

(b) $\vec{\rho} = \dfrac{\vec{R}}{2} + \dfrac{2\vec{S}}{3}$

ANSWER:

(a) $\vec{A} = (2, 3, 4, 5, 6) - (0, 1, 3, 4, 7) = (2, 2, 1, 1, -1)$

(b) $\vec{\rho} = (1, 3/2, 2, 5/2, 3) + (0, 2/3, 2, 8/3, 14/3) = (1, 13/6, 4, 31/6, 23/3)$

9. Use $\vec{\alpha} = (1.1, 3.7, 4.1, 5.6)$ and $\vec{\beta} = (10.5, 10.7, 10.8, 11.1)$ to find the following vectors.

    (a) $\vec{Z} = 2\vec{\alpha} + 3\vec{\beta}$ **(b)** $\vec{\gamma} = 5.2(\vec{\beta} - \vec{\alpha})$

    ANSWER:

    (a) $\vec{Z} = (2.2, 7.4, 8.2, 11.2) + (31.5, 32.1, 32.4, 33.3) = (33.7, 39.5, 40.6, 44.5)$

    (b) $\vec{\gamma} = 5.2(9.4, 7, 6.7, 5.5) = (48.88, 36.4, 34.84, 28.6)$

For Exercises 10–11, a retailer's total monthly sales of three different models of televisions is given by the vector

$$\vec{S} = (17, 23, 8).$$

10. If the sales for each model goes up by 5 the next month, what is $\vec{Q}$, the next month's total sales?

    ANSWER:

    $\vec{Q} = (22, 28, 13)$

11. If the next month's sales for each model are down 10%, what is $\vec{R}$, the next month's total sales? Round to the nearest whole number.

    ANSWER:

    $\vec{R} = (15, 21, 7)$

For Exercises 12–15 perform the operations on the given 3-dimensional vectors.

$$\vec{q} = 3\vec{i} + \vec{j} \qquad \vec{r} = \vec{i} + 2\vec{j} - \vec{k} \qquad \vec{s} = -\vec{i} + 3\vec{j} + 2\vec{k} \qquad \vec{t} = -2\vec{j} + 3\vec{k}$$

12. $\vec{q} \cdot \vec{r}$

    ANSWER:

    $\vec{q} \cdot \vec{r} = 3(1) + 1(2) + 0(-1) = 5$

13. $\vec{s} \cdot \vec{t}$

    ANSWER:

    $\vec{s} \cdot \vec{t} = -1(0) + 3(-2) + 2(3) = 0$

14. $(\vec{r} \cdot \vec{r})\vec{s}$

    ANSWER:

$$(\vec{r} \cdot \vec{r})\vec{s} = (1(1) + 2(2) - 1(-1))\vec{s}$$
$$= 6\vec{s}$$
$$= -6\vec{i} + 18\vec{j} + 12\vec{k}$$

15. $((\vec{r} \cdot \vec{q})\vec{r}) \cdot \vec{q}$

    ANSWER:

$$((\vec{r} \cdot \vec{q})\vec{r}) \cdot \vec{q} = ((1(3) + 2(1) - 1(0))\,\vec{r}) \cdot \vec{q}$$
$$= (5\vec{i} + 10\vec{j} - 5\vec{k}) \cdot \vec{q}$$
$$= (5(3) + 10(1) - 5(0)) = 25$$

Problems

16. A particle is acted on by two forces, one of them to the west and of magnitude 1 dyne, and the other in the direction $60°$ north of east and of magnitude 2 dynes. What third force acting on the particle would keep it in equilibrium? That is, what third force would make the sum of all three forces zero? (A dyne is a unit of force.)

    ANSWER:

    The third force must have same magnitude with the sum of first two forces but opposite direction. Suppose that the forces given are $\vec{F}_1$ and $\vec{F}_2$ and that $\vec{F}_3$ is the force we want. Using the Law of Cosines in Figure 10.7, we have

$$\|\vec{F}_3\|^2 = 1^2 + 2^2 - 2 \cdot 1 \cdot 2 \cos 60° = 1 + 4 - 2 = 3$$
$$\|\vec{F}_3\| = \sqrt{3}.$$

To find the direction of $\vec{F}_3$, use the Law of Sines to find angle θ:

$$\frac{\sin 60°}{\sqrt{3}} = \frac{\sin \theta}{2}$$

$$\sin \theta = \frac{2}{\sqrt{3}} \sin 60° = 1.$$

Thus, $\theta = 90°$.

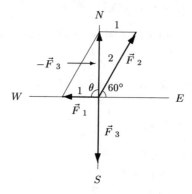

Figure 10.7

Alternatively, notice that since

$$2^2 = 1 + (\sqrt{3})^2,$$

the triangle with $\vec{F}_1$, $\vec{F}_2$, $-\vec{F}_3$ as sides must be a right triangle. So a force of magnitude $\sqrt{3}$ dynes pointing due south keeps the particle equilibrium.

17. A cyclist goes at 6 mph due north and feels the wind coming against him with a relative velocity of 4 mph due west. What is the actual velocity of the wind?

ANSWER:

The actual velocity of the wind, $\vec{w}$, is the sum of the cylist's velocity and the relative wind velocity he feels. See Figure 10.8. Using Pythagoras' Theorem, we have

$$\|\vec{w}\|^2 = 4^2 + 6^2 = 52, \quad \text{so} \quad \|\vec{w}\| = \sqrt{52}.$$

In addition

$$\theta = \arctan\left(\frac{2}{3}\right) = 33.7°.$$

So the actual velocity of the wind is $\sqrt{52} \approx 7.21$ mph at an angle of due $33.7°$ west of north.

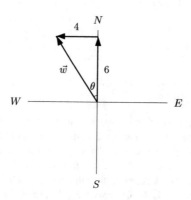

Figure 10.8

18. A gun mounted on a train points vertically upward. The train moves horizontally due east at 80 mph and the gun fires a bullet with a muzzle velocity of 80 mph. What is the speed and direction of the bullet relative to the ground?

ANSWER:

By the Pythagorean Theorem, the velocity of the bullet relative to the ground is $80\sqrt{2}$ mph toward the east at $45°$ to the horizontal.

19. In Figure 10.9, each grid square is 1 unit along each side.

(a) Write $\vec{v}$, $\vec{w}$, and $\vec{v} + \vec{w}$ in component form.

(b) Find a vector perpendicular to the displacement vector $\overrightarrow{RS}$.

(c) What is the angle between the x-axis and the vector $\vec{v}$?

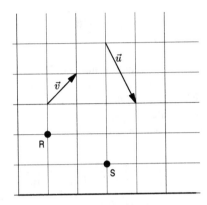

Figure 10.9

ANSWER:

(a) $\vec{v} = \vec{i} + \vec{j}$, $\vec{w} = \vec{i} - 2\vec{j}$, and $\vec{v} + \vec{w} = 2\vec{i} - \vec{j}$

(b) The vector $\overrightarrow{RS}$ can be written $2\vec{i} - \vec{j}$. Thus, a perpendicular vector would be $\vec{i} + 2\vec{j}$.

(c) If θ is the angle between the x-axis and the vector $\vec{v}$, then

$$\cos\theta = \frac{1}{\|\vec{v}\|} = \frac{1}{\sqrt{2}} = \frac{\sqrt{2}}{2}$$

Thus, $\theta = 45°$.

20. A particle in equilibrium is acted on by three forces, two of which have components $6\vec{i} + 5\vec{j} - 2\vec{k}$ and $3\vec{i} - 10\vec{j} + 8\vec{k}$ respectively. What are the components of the third force? (For equilibrium, the sum of the three forces must be zero.)

ANSWER:

The third force must have the same magnitude as the sum of first two forces and be in the opposite direction. Since

$$(6\vec{i} + 5\vec{j} - 2\vec{k}) + (3\vec{i} - 10\vec{j} + 8\vec{k}) = 9\vec{i} - 5\vec{j} + 6\vec{k},$$

so the third force has components

$$-(9\vec{i} - 5\vec{j} + 6\vec{k}) = -9\vec{i} + 5\vec{j} - 6\vec{k}.$$

21. Show that as t changes, the tip of the vector $\vec{r} = (1 + 2t)\vec{i} + (3 - 4t)\vec{j}$ moves along a straight line.

ANSWER:

We rewrite the vector as follows

$$\vec{r} = (1 + 2t)\vec{i} + (3 - 4t)\vec{j} = (\vec{i} + 3\vec{j}) + (2\vec{i} - 4\vec{j})t.$$

Thus, we see that $\vec{r}$ is the sum of the vector $\vec{i} + 3\vec{j}$ and a scalar multiple of the vector $2\vec{i} - 4\vec{j}$. See Figure 10.10. As t changes, the tip of $\vec{r}$ moves along the line through $(1, 3)$ and parallel to the vector $2\vec{i} - 4\vec{j}$.

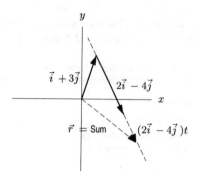

Figure 10.10

22. There are six people taking a vocational exam with both written and oral parts. Their scores on the written section (out of 100) are given by the vector $\vec{r} = (71, 49, 87, 98, 81, 63)$. Their scores on the oral section (out of 100) are given by the vector $\vec{s} = (79, 73, 82, 95, 97, 66)$. Find a vector giving their composite scores if

   (a) the written part counts twice as much as the oral part.
   (b) the oral part counts twice as much as the written part.

   ANSWER:

   (a) The vector is given by $2\vec{r} + \vec{s}$, which is $(221, 171, 256, 291, 259, 192)$.
   (b) The vector is given by $\vec{r} + 2\vec{s}$, which is $(229, 195, 251, 288, 275, 195)$.

23. The revenue earned, in millions of dollars, by a movie from domestic and overseas box-office sales is given by a vector of the form $\vec{r} = (\text{Domestic}, \text{Overseas})$. The top three grossing movies in 1998, *Armageddon*, *Saving Private Ryan*, and *Godzilla*, have revenue vectors $\vec{r}_A = (201.6, 347.6)$, $\vec{r}_S = (215.9, 256.9)$, $\vec{r}_G = (136.3, 239.7)$, respectively.

   (a) What does the vector $\vec{r}_A$ tell you?
   (b) What was the total amount earned by *Saving Private Ryan* at the box office?
   (c) What was the total amount earned by all three at the domestic box office?

   ANSWER:

   (a) The movie *Armageddon* earned $201.6 million at the domestic (US) box office and $347.6 million overseas.
   (b) The total amount earned by *Saving Private Ryan* is $215.9 + 256.9 = \$472.7$ million.
   (c) The total amount earned domestically by the three movies is $201.6 + 215.9 + 136.3 = \$553.8$ million.

24. An airplane is flying at an airspeed of 650 km/hr in a crosswind blowing from the southeast at a speed of 60 km/hr.

   (a) In what direction should the plane head to end up going east? What will its speed be relative to the ground?
   (b) If the same crosswind is blowing on the return trip, in what direction should the plane head to end up going west? What will its speed be relative to the ground then?

   ANSWER:

   (a) Figure 10.11 shows the vectors of the plane and the crosswind. The crosswind is blowing at an angle of 45 deg west of north, so its vector can be broken into components $-30\sqrt{2}\vec{i} + 30\sqrt{2}\vec{j}$. Using this, we see that

$$\sin\theta = \frac{30\sqrt{2}}{650} = 0.06527$$

$$\theta = \arcsin(0.06527) = 3.74°.$$

Thus, $\theta = 3.74°$ south of east. To find x, the speed of the airplane relative to the ground, solve

$$\cos 3.74° = \frac{x + 30\sqrt{2}}{650}$$

$$0.99787 = \frac{x + 30\sqrt{2}}{650}$$

$$648.6155 = x + 30\sqrt{2}$$

$$x = 606.19 \text{ km/hr.}$$

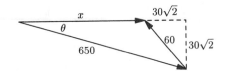

Figure 10.11

(b) Figure 10.12 shows the vectors of the plane and the crosswind for the return trip. The crosswind vector can again be broken into components $-30\sqrt{2}\vec{i} + 30\sqrt{2}\vec{j}$, and we get the same angle θ. In this case, however, $\theta = 3.74°$ south of west. To find y, the speed of the airplane relative to the ground, solve

$$\cos 3.74° = \frac{y - 30\sqrt{2}}{650}$$

$$0.99787 = \frac{y - 30\sqrt{2}}{650}$$

$$648.6155 = y - 30\sqrt{2}$$

$$y = 691.04 \text{ km/hr.}$$

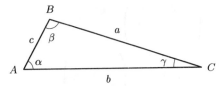

Figure 10.12

25. In Figure 10.13, use the dot product to prove the Law of Cosines:

$$a^2 = b^2 + c^2 - 2bc\cos\alpha.$$

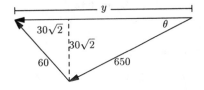

Figure 10.13

ANSWER:

$$a^2 = ||\overrightarrow{BC}||^2 = \overrightarrow{BC} \cdot \overrightarrow{BC} = (\overrightarrow{AC} - \overrightarrow{AB}) \cdot (\overrightarrow{AC} - \overrightarrow{AB})$$
$$= \overrightarrow{AC} \cdot \overrightarrow{AC} - 2\overrightarrow{AB} \cdot \overrightarrow{AC} + \overrightarrow{AB} \cdot \overrightarrow{AB}$$
$$= ||\overrightarrow{AC}||^2 + ||\overrightarrow{AB}||^2 - 2(||\overrightarrow{AB}||||\overrightarrow{AC}||)\cos\alpha$$
$$= b^2 + c^2 - 2bc\cos\alpha$$

Therefore $a^2 = b^2 + c^2 - 2bc\cos\alpha$.

26. For what value of x are $3\vec{i} - 2\vec{j} + 3\vec{k}$ and $x\vec{i} - 2\vec{j} + 2\vec{k}$ perpendicular?

ANSWER:

The dot product should be zero,

$$(3\vec{i} - 2\vec{j} + 3\vec{k}) \cdot (x\vec{i} - 2\vec{j} + 2\vec{k}) = 3x + 4 + 6 = 0,$$

so $x = -10/3$.

27. How much work is done pushing a 20 lb stroller with a 35 lb toddler in it up a 50 yard hill if the slope of the hill is $10°$? What if the slope is $20°$?

ANSWER:

We use Work= $\vec{F} \cdot \vec{d}$. Convert 50 yds to 150 ft. As seen in Figure 10.14, $\vec{F}$ is the total weight on the stroller and the child (55 lb), and $\|\vec{d}\|$ is given by $150 \cos 80° = 26.05$. Thus, Work= $55(26.05) = 1432.75$ ft-lbs. If the slope is $20°$, then $\|\vec{d}\| = 150 \cos 70° = 51.30$. Thus, Work $= 55(51.30) = 2821.5$ ft-lbs.

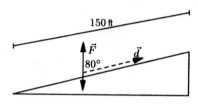

Figure 10.14

28. Evaluate the following expressions given that

$$A = \begin{pmatrix} 1 & 3 & 2 \\ 4 & -1 & -3 \\ 4 & 0 & 2 \end{pmatrix} \qquad B = \begin{pmatrix} 5 & 1 & -3 \\ 2 & -4 & 0 \\ -1 & -2 & 3 \end{pmatrix}$$

(a) $3A$

(b) $-2B$

(c) $A - B$

(d) $3A - 2B$

(e) $-4A + 4B$

(f) $((A - B) - B) - B$

ANSWER:

(a) $3A = 3 \begin{pmatrix} 1 & 3 & 2 \\ 4 & -1 & -3 \\ 4 & 0 & 2 \end{pmatrix} = \begin{pmatrix} 3 & 9 & 6 \\ 12 & -3 & -9 \\ 12 & 0 & 6 \end{pmatrix}$

(b) $-2B = -2 \begin{pmatrix} 5 & 1 & -3 \\ 2 & -4 & 0 \\ -1 & -2 & 3 \end{pmatrix} = \begin{pmatrix} -10 & -2 & 6 \\ -4 & 8 & 0 \\ 2 & 4 & -6 \end{pmatrix}$

(c) $A - B = \begin{pmatrix} 1 & 3 & 2 \\ 4 & -1 & -3 \\ 4 & 0 & 2 \end{pmatrix} - \begin{pmatrix} 5 & 1 & -3 \\ 2 & -4 & 0 \\ -1 & -2 & 3 \end{pmatrix} = \begin{pmatrix} -4 & 2 & 5 \\ 2 & 3 & -3 \\ 5 & 2 & -1 \end{pmatrix}$

(d) $3A - 2B = \begin{pmatrix} 3 & 9 & 6 \\ 12 & -3 & -9 \\ 12 & 0 & 6 \end{pmatrix} + \begin{pmatrix} -10 & -2 & 6 \\ -4 & 8 & 0 \\ 2 & 4 & -6 \end{pmatrix} = \begin{pmatrix} -7 & 7 & 12 \\ 8 & 5 & -9 \\ 14 & 4 & 0 \end{pmatrix}$

(e) $-4A + 4B = -4 \begin{pmatrix} 1 & 3 & 2 \\ 4 & -1 & -3 \\ 4 & 0 & 2 \end{pmatrix} + 4 \begin{pmatrix} 5 & 1 & -3 \\ 2 & -4 & 0 \\ -1 & -2 & 3 \end{pmatrix} = \begin{pmatrix} 16 & -8 & -20 \\ -8 & -12 & 12 \\ -20 & -8 & 4 \end{pmatrix}$

(f) $((A - B) - B) - B = A - 3B = \begin{pmatrix} 1 & 3 & 2 \\ 4 & -1 & -3 \\ 4 & 0 & 2 \end{pmatrix} - 3 \begin{pmatrix} 5 & 1 & -3 \\ 2 & -4 & 0 \\ -1 & -2 & 3 \end{pmatrix} = \begin{pmatrix} -14 & 0 & 11 \\ -2 & 11 & -3 \\ 7 & 6 & -7 \end{pmatrix}$

29. Evaluate the following expressions given that

$$\mathbf{R} = \begin{pmatrix} 2 & -4 \\ 1 & 5 \end{pmatrix} \qquad \mathbf{S} = \begin{pmatrix} 1 & 3 \\ -2 & 7 \end{pmatrix} \qquad \vec{u} = (1,3) \qquad \vec{v} = (-1,2)$$

(a) $\mathbf{R}\vec{u}$ **(b)** $\mathbf{S}(\vec{u} + \vec{v})$ **(c)** $(\mathbf{R} - \mathbf{S})\vec{v}$ **(d)** $(\vec{u} \cdot \vec{v})(\mathbf{R} + \mathbf{S})$

ANSWER:

(a) $\mathbf{R}\vec{u} = \begin{pmatrix} 2 & -4 \\ 1 & 5 \end{pmatrix}(1,3) = (2(1) - 4(3), 1(1) + 5(3)) = (-10, 16)$

(b) $\mathbf{S}(\vec{u} + \vec{v}) = \begin{pmatrix} 1 & 3 \\ -2 & 7 \end{pmatrix}(0,5) = (1(0) + 3(5), -2(0) + 7(5)) = (15, 35)$

(c) $(\mathbf{R} - \mathbf{S})\vec{v} = \begin{pmatrix} 1 & -7 \\ 3 & -2 \end{pmatrix}(-1,2) = (1(-1) - 7(2), 3(-1) - 2(2)) = (-15, -7)$

(d) $(\vec{u} \cdot \vec{v})(\mathbf{R} + \mathbf{S}) = 5\begin{pmatrix} 3 & -1 \\ -1 & 12 \end{pmatrix} = \begin{pmatrix} 15 & -5 \\ -5 & 60 \end{pmatrix}$

30. A country has two main political parties. The vector $\vec{P} = (f, s, n)$ gives the number of people who are members of the first party, the second party, and neither party, respectively.

(a) Suppose $\vec{P}_{new} = \mathbf{T}\vec{P}_{old} = \begin{pmatrix} 0.95 & 0.12 & 0.10 \\ 0.04 & 0.85 & 0.15 \\ 0.01 & 0.03 & 0.75 \end{pmatrix}\vec{P}_{old}$. Describe in words what this tells you about the loyalties of members to their parties. Be specific.

(b) Let $\vec{P}_{2003} = (30, 27, 15)$ be the population vector (in tens of thousands) in the year 2003. Evaluate $\vec{P}_{2004}$ and $\vec{P}_{2005}$.

ANSWER:

(a) Members of the first party are quite loyal: 95% of them remain members of that party from one year to the next. Of the remaining 5%, 4% switch to the second party and 1% become members of neither party. Members of the second party are not as loyal: 85% of them remain members of that party from one year to the next. Of the remaining 15%, 12% switch to the second party and 3% become members of neither party.

(b)

$$\vec{P}_{2004} = \begin{pmatrix} 0.95 & 0.12 & 0.10 \\ 0.04 & 0.85 & 0.15 \\ 0.01 & 0.03 & 0.75 \end{pmatrix}(30, 27, 15) = (33.24, 26.4, 12.36)$$

$$\vec{P}_{2005} = \begin{pmatrix} 0.95 & 0.12 & 0.10 \\ 0.04 & 0.85 & 0.15 \\ 0.01 & 0.03 & 0.75 \end{pmatrix}(33.24, 26.4, 12.36) = (35.982, 25.6236, 10.3944)$$

31. The inverse of a matrix $\mathbf{A}$, denoted $\mathbf{A}^{-1}$, is such that if $\vec{v} = \mathbf{A}\vec{u}$, then $\vec{u} = \mathbf{A}^{-1}\vec{v}$. For a matrix $\mathbf{A} = \begin{pmatrix} a & b \\ c & d \end{pmatrix}$, the inverse $\mathbf{A}^{-1}$ is given by $\mathbf{A}^{-1} = \dfrac{1}{D}\begin{pmatrix} d & -b \\ -c & a \end{pmatrix}$, where $D = ad - bc$. $\mathbf{A}^{-1}$ is undefined if $D = 0$.

(a) Let $\mathbf{A} = \begin{pmatrix} 1 & -3 \\ -2 & 5 \end{pmatrix}$. Find $\mathbf{A}^{-1}$ and verify that if $\vec{v} = \mathbf{A}\vec{u}$, then $\vec{u} = \mathbf{A}^{-1}\vec{v}$ for $\vec{u} = (x, y)$.

(b) The formula for $\mathbf{A}^{-1}$ can be verified algebraically. Suppose $\vec{v} = \mathbf{A}\vec{u}$. Then $\mathbf{A}^{-1}\vec{v} = \mathbf{A}^{-1}(\mathbf{A}\vec{u})$. Show that $\mathbf{A}^{-1}(\mathbf{A}\vec{u}) = \vec{u}$ for $\mathbf{A} = \begin{pmatrix} a & b \\ c & d \end{pmatrix}$ and $\vec{u} = (x, y)$.

ANSWER:

(a) $\mathbf{A}^{-1} = \dfrac{1}{5 - 6}\begin{pmatrix} 5 & 3 \\ 2 & 1 \end{pmatrix} = -\begin{pmatrix} 5 & 3 \\ 2 & 1 \end{pmatrix} = \begin{pmatrix} -5 & -3 \\ -2 & -1 \end{pmatrix}$. Next compute $\mathbf{A}\vec{u}$:

$$\begin{pmatrix} 1 & -3 \\ -2 & 5 \end{pmatrix}(x, y) = (x - 3y, -2x + 5y).$$

Now compute $\mathbf{A}^{-1}(x - 3y, -2x + 5y)$:

$$\begin{pmatrix} -5 & -3 \\ -2 & -1 \end{pmatrix}(x - 3y, -2x + 5y) = (-5(x - 3y) - 3(-2x + 5y), -2(x - 3y) - 1(-2x + 5y))$$

$$= (x, y)$$

$$= \vec{u}.$$

(b) Compute $\mathbf{A}^{-1}(\mathbf{A}\vec{u})$:

$$\mathbf{A}^{-1}(\mathbf{A}\vec{u}) = \mathbf{A}^{-1}\begin{pmatrix} a & b \\ c & d \end{pmatrix}(\mathbf{x}, \mathbf{y})$$

$$= \frac{1}{ad - bc}\begin{pmatrix} d & -b \\ -c & a \end{pmatrix}(ax + by, cx + dy)$$

$$= \frac{1}{ad - bc}(d(ax + by) - b(cx + dy), -c(ax + by) + a(cx + dy))$$

$$= \frac{1}{ad - bc}((ad - bc)x, (ad - bc)y)$$

$$= (x, y)$$

$$= \vec{u}.$$

Chapter 11 Exam Questions

Exercises

Are the sequences in Exercises 1–4 arithmetic, geometric, or neither? If arithmetic or geometric, give a formula for the nth term.

1. $1, 4, 7, 10, \ldots$
   ANSWER:
   Arithmetic, $a_n = 1 + (n-1)3 = -2 + 3n$

2. $1, 2, 4, 7, \ldots$
   ANSWER:
   Neither

3. $1, \frac{1}{3}, \frac{1}{9}, \frac{1}{27}, \ldots$
   ANSWER:
   Geometric, $a_n = \dfrac{1}{3^{n-1}}$

4. $1, \frac{1}{2}, 0, -\frac{1}{2}, \ldots$
   ANSWER:
   Arithmetic, $a_n = 1 - \dfrac{1}{2}(n-1) = \dfrac{3}{2} - \dfrac{n}{2}$

5. Complete the table with the terms of the arithmetic sequence $a_1, a_2, \ldots, a_n$, and the sequence of partial sums, $S_1, S_2, \ldots, S_n$. Find d, where $a_n = a_1 + (n-1)d$.

Table 11.1

n	1	2	3	4	5	6	7	8
a_n	1	4						
S_n	1							

ANSWER:
$d = 3$

Table 11.2

n	1	2	3	4	5	6	7	8
a_n	1	4	7	10	13	16	19	22
S_n	1	5	12	22	35	51	70	92

6. Expand the following sum. (Do not evaluate.)
$$\sum_{n=3}^{11} (-1)^{n-1} n^2$$

ANSWER:

$$\sum_{n=3}^{11} (-1)^{n-1} n^2 = 9 - 16 + 25 - 36 + 49 - 64 + 81 - 100 + 121$$

7. Write the following sum using sigma notation.
$$3 + 5 + 7 + 8 + 11 + 13$$

ANSWER:

$$3 + 5 + 7 + 8 + 11 + 13 = \sum_{n=1}^{6} 1 + 2n$$

8. Without a calculator, find the following sum

$$\sum_{k=1}^{40} (2k + 1)$$

ANSWER:

$$\sum_{k=1}^{40} (2k + 1) = 2 \sum_{k=1}^{40} k + \sum_{k=1}^{40} 1$$
$$= 2 \left(\frac{1}{2}\right)(40)(41) + 40 = 1680.$$

9. Decide whether or not each of the following is a geometric series. If it is, give the ratio between terms. If not, explain why.
   (a) $1 - 2 + 4 - 8 + 16 - \cdots + 256$
   (b) $1 - 2 + 3 - 4 + 5 - \cdots + 9$
   ANSWER:
   (a) Yes, $r = -2$.
   (b) No, ratio between successive terms is not constant.

10. Find the sum of the series

$$\sum_{n=0}^{6} 4 \left(\frac{1}{3}\right)^n$$

ANSWER:

$$\sum_{n=0}^{6} 4 \left(\frac{1}{3}\right)^n = \frac{4 \left(1 - \left(\frac{1}{3}\right)^7\right)}{1 - \frac{1}{3}}$$
$$= \frac{4372}{729}$$

11. Write the following sum using sigma notation

$$2 + 6 + 18 + 54 + 162 + 486$$

ANSWER:

$$\sum_{n=0}^{5} 2(3)^n$$

12. Identify each series as arithmetic, geometric, or neither.
   (a) $1^2 + 2^2 + 3^2 + 4^2 + \ldots$
   (b) $3.0 + 4.6 + 6.2 + 7.8 + \ldots$
   (c) $\frac{9}{4} + \frac{3}{2} + 1 + \frac{2}{3} + \ldots$
   (d) $\frac{1}{2} + \frac{5}{4} + 2 + \frac{11}{4} + \ldots$
   ANSWER:

   (a) Neither
   (b) Arithmetic
   (c) Geometric
   (d) Arithmetic

13. Decide whether or not each of the following is a geometric series. If it is, give the ratio between terms. If not, explain why.

(a) $x + \frac{x^2}{2} + \frac{x^3}{3} + \frac{x^4}{4} + \cdots$

(b) $x + \frac{x^2}{2} + \frac{x^3}{4} + \frac{x^4}{8} + \cdots$

ANSWER:

(a) No, ratio between successive terms is not constant.

(b) Yes, $r = \frac{x}{2}$

14. Find the sum of the series

$$3 + \frac{3}{2} + \frac{3}{4} + \frac{3}{8} + \cdots$$

ANSWER:

$$3 + \frac{3}{2} + \frac{3}{4} + \frac{3}{8} + \cdots = \frac{3}{1 - \frac{1}{2}} = 6$$

15. Find the sum of the series

$$\sum_{k=0}^{\infty} \frac{2^k + 3}{3^k}$$

ANSWER:

$$\sum_{k=0}^{\infty} \frac{2^k + 3}{3^k} = \sum_{k=0}^{\infty} \left(\frac{2}{3}\right)^k + \sum_{k=0}^{\infty} 3 \left(\frac{1}{3}\right)^k$$
$$= \frac{1}{1 - \frac{2}{3}} + \frac{3}{1 - \frac{1}{3}}$$
$$= 3 + \frac{9}{2} = 7.5$$

Problems

16. Decide which of the following graphs represents an arithmetic sequence and which represents a geometric sequence. For the arithmetic sequence, estimate the common difference, d. For the geometric sequence, estimate the common ratio, r.

(a)

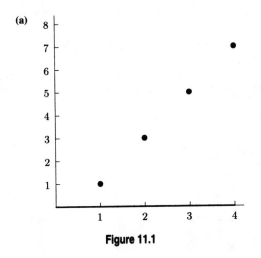

Figure 11.1

(b)

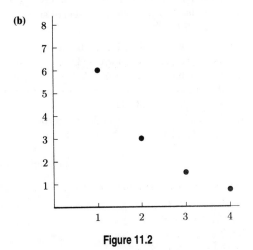

Figure 11.2

ANSWER:

(a) Arithmetic, $d = 2$.

(b) Geometric, $r = 1/2$.

17. A new employee is offered two salary packages. Bost start with an initial annual salary of $48,000$. Package A offers a $2000 salary increase each year. Package B offers a 3.5% salary increase each year.

    (a) Write the first four terms of the sequence, a_n, of yearly salaries and give a formula for the general term for each salary package. (Round to the nearest whole dollar)

    (b) After how many years will Package B pay a higher annual salary than Package A? If the employee plans to work for the company for exactly this many years, which salary package would you recommend? Why?

    ANSWER:

    (a) For Package A : $48,000$, $50,000$, $52,000$, $54,000$. The formula for the general term is

$$a_n = 48,000 + 2000(n-1) = 46000 + 2000n.$$

    For Package B : $48,000$, $49,680$, $51,419$, $53,219$. The formula for the general term is

$$a_n = 48,000(1.035)^{n-1}.$$

    (b) Either graph both functions or continue finding values.
    For Package A, $a_{11} = 68,000$ and $a_{12} = 70,000$.
    For Package B, $a_{11} = 67,709$ and $a_{12} = 70,078$.
    Thus, Package B pays a higher annual salary than Package A after 12 years. If the employee were to work exactly 12 years, Package A would still be better overall because the cumulative salaries for the 12 years would still be higher than the cumulative salaries for Package B.

18. A sequence a_n can be defined by a recurrence relation, which gives the first term, a_1, and a formula for finding a_n in terms of the previous term a_{n-1}. For the following recurrence relation, find the first four terms of the sequence and give a formula for the general term.

$$a_1 = 1; \quad a_n = \frac{a_{n-1}}{2}$$

    ANSWER:

$$a_1 = 1, \ a_2 = \frac{1}{2}, \ a_3 = \frac{1}{4}, \ a_4 = \frac{1}{8}; a_n = \left(\frac{1}{2}\right)^{n-1}$$

19. (a) Consider the sequence $1^2, 2^2, 3^2, 4^2, \ldots$. Is this an arithmetic sequence? Why or why not?

    (b) Find the sums of the following arithmetic series involving the odd integers: $1 + 3$, $1 + 3 + 5$, $1 + 3 + 5 + 7$. What do you notice about the sums of these series and the sequence from part (a)?

    (c) Use what you know about the sum of an arithmetic series to show that the sum of the first n odd numbers equals n^2.

    ANSWER:

    (a) The sequence $1, 4, 9, 16, \ldots$, is not an arithmetic sequence because the terms do not increase by a constant amount d. Rather the terms in increase by an increasing amount: $3, 5, 7, \ldots$

    (b) We see that $1 + 3 = 4$, $1 + 3 + 5 = 9$, and $1 + 3 + 5 + 7 = 16$. The sums of these series of odd numbers give the terms of the sequence of squares from part (a). In other words, it seems that each perfect square is the sum of an arithmetic series of odd numbers.

    (c) We need to find the sum of the arithmetic series $1 + 3 + 5 + \cdots + a_n$, where a_n is the nth odd number. For this series, $a_1 = 1$ and $d = 2$, and so we have

$$
\begin{aligned}
\text{Sum of first } n \text{ odd numbers} &= \frac{1}{2}n(2a_1 + (n-1)d) \\
&= \frac{1}{2}n(2 \cdot 1 + (n-1)2) \\
&= \frac{1}{2}n(2 + 2n - 2) \\
&= \frac{1}{2}n(2n) = n^2.
\end{aligned}
$$

Thus, the sum of the first n odd numbers equals n^2.

20. Assume a_1 and a_2 are constants. The symbol $\sum_{i=1}^{2}(x-a_i)^2$ is a sum, and its value is given by

$$\sum_{i=1}^{2}(x-a_i)^2 = (x-a_1)^2 + (x-a_2)^2.$$

Let

$$f(x) = \sum_{i=1}^{2}(x-a_i)^2.$$

(a) If $a_1 = a_2 < 0$, does the equation $f(x) = 0$ have a solution? If so, what is it?

(b) If $a_1 \neq a_2$, does the equation $f(x) = 0$ have a solution? If so, what is it? Explain your answer.

(c) If $a_2 = 0$, solve for x: $f(x) = 2a_1{}^2$. (Your answer will contain a_1.)

ANSWER:

(a) If $a_1 = a_2$, then $f(x) = 2(x-a_1)^2 = 0$ has one solution at $x = a_1$

(b) If $a_1 \neq a_2$, then $f(x) = (x-a_1)^2 + (x-a_2)^2$ and the terms $(x-a_1)^2$ and $(x-a_2)^2$ are both nonnegative and they are not zero for the same value of x. Therefore $f(x) = 0$ has no solution.

(c) If $a_2 = 0$ then $f(x) = (x-a_1)^2 + (x-0)^2 = (x-a_1)^2 + x^2$. Solving $f(x) = 2a_1^2$, gives

$$(x-a_1)^2 + x^2 = 2a_1^2$$
$$x^2 - 2a_1 x + a_1^2 + x^2 = 2a_1^2$$
$$2x^2 - 2a_1 x - a_1^2 = 0$$

so $$x = \frac{2a_1 \pm \sqrt{4a_1^2 + 8a_1^2}}{4} = \frac{2a_1 \pm \sqrt{12a_1^2}}{4}$$

$$x = \frac{2a_1 \pm 2a_1\sqrt{3}}{4} = \frac{a_1(1 \pm \sqrt{3})}{2}$$

21. A person decides to walk for 10 minutes one day, and then increase his walks by 2 minutes each day for a month.

(a) Find the first seven terms of the arithmetic sequence $a_1, a_2, \ldots, a_n$, where a_n is the length of time he walks on day n.

(b) Find and interpret the partial sums $S_1, S_2, \ldots, S_7$.

(c) Find formulas for a_n and S_n.

ANSWER:

(a) 10, 12, 14, 16, 18, 20, 22

(b) 10, 22, 36, 52, 70, 90, 112; these represent the total time walked after n days.

(c)
$$a_n = 10 + (n-1)2 = 8 + 2n$$

$$S_n = \frac{1}{2}n(2(10) + (n-1)2)$$
$$= \frac{n}{2}(18 + 2n)$$
$$= 9n + n^2$$

22. (a) Use summation notation to denote the series $8(0.8)^5 + 8(0.8)^6 + \ldots + 8(0.8)^{20}$.

(b) Write an expression for the sum of the series, then evaluate the sum, rounded to three decimal places.

ANSWER:

(a) $\sum_{k=5}^{20} 8(0.8)^k$

(b) $\frac{8(0.8^5 - 0.8^{21})}{0.2} \approx 12.738$

23. To save for their child's college education, parents put $100 a month into a certificate of deposit that pays 3% annual interest, compounded monthly.

(a) How much will they have saved if they do this for 18 years?

(b) Repeat the problem if the interest rate is 6%.

(c) How much more money per month would they need to put in at 3% to equal the amount they would have had saving $100 per month at 6%?

ANSWER:

(a) Each month, the account will earn $3/12 = 0.25\%$ interest. Thus,

$$B_n = 100 + 100(1.0025) + 100(1.0025)^2 + \cdots 100(1.0025)^{n-1}.$$

The formula for the sum is given by

$$B_n = \frac{100(1 - 1.0025^n)}{1 - 1.0025}$$

Since 18 years=216 months, we get

$$B_{216} = \frac{100(1 - 1.0025^{216})}{1 - 1.0025} = \$28,594.03$$

(b) Using a similar formula with a monthly rate of $6/12 = .05\%$,

$$B_{216} = \frac{100(1 - 1.005^{216})}{1 - 1.005} = \$38,735.32$$

(c) We need to solve the following for P:

$$\frac{P(1 - 1.0025^{216})}{1 - 1.0025} = 38,735.32$$

$$P = \$135.47$$

24. A patient takes 500 mg of a pain-killing drug twice a day. Every twelve hours, the patient's body metabolizes 30% of the drug present.

(a) Write a geometric series that gives the drug level in the patient's body after the nth dose.
(b) What quantity of the drug remains after one week (right after the 14th dose)?
(c) If the patient stops taking the drug after one week, what quantity of the drug remains one day later? One week later?

ANSWER:

(a) Since 70% of the drug remains in the body after 12 hours, we get

$$Q_n = 500 + 500(0.7) + 500(0.7)^2 + \cdots + 500(0.7)^{n-1}.$$

(b)

$$Q_{14} = \frac{500(1 - 0.7^{14})}{1 - 0.7} = 1655.363 \text{ mg}$$

(c) Since two 12-hour periods have gone by, the amount remaining is given by $(1655.363)(0.7)^2 = 811.128$ mg. One week later, the amount is given by $(1655.363)(0.7)^{14} = 5.501$ mg.

25. (a) Find the sum of the series $1 - \frac{x}{2} + \frac{x^2}{4} - \frac{x^3}{8} + \ldots$ for $|x| \leq 2$.
(b) Use your answer to part (a) to evaluate $1 - \frac{1}{2} + \frac{1}{4} - \frac{1}{8} + \ldots$.

ANSWER:

(a) Rewrite the series as

$$1 + \left(-\frac{x}{2}\right) + \left(-\frac{x}{2}\right)^2 + \left(-\frac{x}{2}\right)^3 + \ldots$$

then the sum, S, is

$$S = \frac{1}{1 - \left(-\frac{x}{2}\right)} = \frac{2}{2 + x}.$$

(b) Let $x = 1$; we have $S = \frac{2}{3}$.

26. **(a)** Subtract the following two equations and solve for S in terms of a and x:

$$S = a + ax + ax^2 + ax^3 + \cdots + ax^{10}$$
$$xS = ax + ax^2 + ax^3 + ax^4 + \cdots + ax^{11}$$

(b) By analogy with part (a), suppose n is a positive integer and that

$$S = a + ax + ax^2 + ax^3 + \cdots + ax^n.$$

Create a second equation analogous to the second one in part (a) then subtract and solve for S.

(c) Suppose $0 < x < 1$. What happens to the value of S in part (b) as $n \to \infty$?

ANSWER:

(a) Subtracting gives

$$(1-x)S = a - ax^{11},$$

so

$$S = \frac{a(1-x^{11})}{1-x}.$$

(b) We subtract

$$S = a + ax + ax^2 + \cdots + ax^n$$
$$xS = ax + ax^2 + ax^3 + \cdots + ax^{n+1}$$
$$(1-x)S = a - ax^{n+1}$$
$$S = \frac{a(1-x^{n+1})}{1-x}.$$

(c) If $0 < x < 1$, then $x^{n+1} \to 0$ as $n \to \infty$, so

$$S \to \frac{a}{1-x}.$$

27. What is the present value of a lottery prize of 1 million dollars, paid out at a rate of $\$20,000$ per year for 50 years? Assume an annual interest rate of 4%.

ANSWER:

$$P = 20,000 + 20,000 \left(\frac{1}{1.04}\right) + 20,000 \left(\frac{1}{1.04}\right)^2 + \cdots + 20,000 \left(\frac{1}{1.04}\right)^{49}$$

$$= \frac{20,000 \left(1 - \left(\frac{1}{1.04}\right)^{50}\right)}{1 - \frac{1}{1.04}}$$

$$= \$446,829.44.$$

28. A patient takes 500 mg of a pain-killing drug twice a day. Every twelve hours, the patient's body metabolizes 30% of the drug present. Assume that the patient continues taking the drug indefinitely. How much of the drug will remain in the patient's body in the long run?

ANSWER:

Since 70% of the drug remains in the body after 12 hours, we get

$$Q = 500 + 500(0.7) + 500(0.7)^2 + \cdots$$

Thus, the effect in the long run is given by

$$Q = \frac{500}{1 - 0.7} = 1666.667 \text{ mg}.$$

Chapter 12 Exam Questions

Exercises

Graph the parametric equations in Exercises 1–3. Assume that the parameter is restricted to values for which the functions are defined. Indicate the direction of the curve. Eliminate the parameter, t, to obtain an equation in terms of y and x.

1. $x = t + 1, \quad y = \sqrt{t}$
   ANSWER:

$$t = x - 1$$
$$y = \sqrt{t}$$
$$ = \sqrt{x - 1}$$

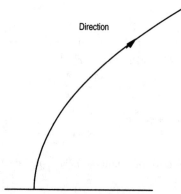

Direction

Figure 12.1

2. $x = e^t, \quad y = t^2$
   ANSWER:

$$x = e^t$$
$$\ln x = t$$
$$y = (\ln x)^2$$

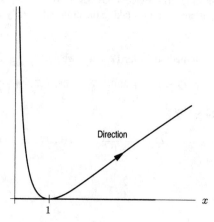

Direction

x

1

Figure 12.2

3. $x = \cos t, \quad y = 2\sin t$
    ANSWER:

$$\cos^2 t + \sin^2 t = 1$$
$$x^2 + \left(\frac{y}{2}\right)^2 = 1$$

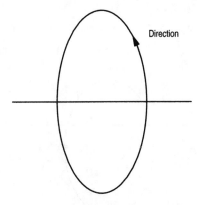

Figure 12.3

4. Describe the motion of a particle whose position at time t is given by $x = f(t)$, $y = g(t)$.

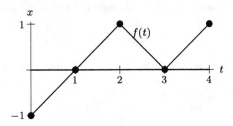

Figure 12.4

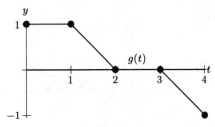

Figure 12.5

    ANSWER:
    Lines from (-1,1) to (0,1) to (1,0) to (0,0) to (1,-1).

5. What are the center and radius of the circle $3(x-1)^2 + 2(y+2)^2 - 12 = -(y-2)^2$.
    ANSWER:

$$3(x-1)^2 + 2(y+2)^2 - 12 = -(y-2)^2$$
$$3(x-1)^2 + 3(y+2)^2 = 12$$
$$(x-1)^2 + (y+2)^2 = 4$$

The center is (1,-2) and the radius is 2.

6. Parameterize a circle with the following properties:

Radius 4, centered at the origin, traversed clockwise starting at (0,4).

ANSWER:

$x = 4\sin t, \quad y = 4\cos t, \quad 0 \le t \le 2\pi$

7. Parameterize a circle with the following properties:

Radius 3, centered at (1,2), traversed counterclockwise starting at (4,2).

ANSWER:

$x = 1 + 3\cos t, \quad y = 2 + 3\sin t, \quad 0 \le t \le 2\pi$

8. For the following ellipse, find the coordinates of the center and the "diameter" in the $x-$ and $y-$directions. Then find an implicit equation for the ellipse.

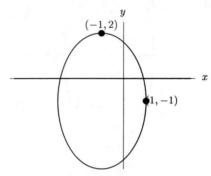

Figure 12.6

ANSWER:

The center is at (-1,-1). The diameter in the $x-$direction is 4 and in the $y-$direction it is 6. An implicit equation would be

$$\frac{(x+1)^2}{4} + \frac{(y+1)^2}{9} = 1.$$

9. Parameterize the ellipse in Exercise 8 traversed in the clockwise direction, starting at the point (-1,2).

ANSWER:

$x = -1 + 2\sin t, \quad y = -1 + 3\cos t, \quad 0 \le t \le 2\pi.$

For the hyperbolas in Exercises 10–11, find

(a) The coordinates of the vertices of the hyperbola and the coordinates of the center.

(b) The equations of the asymptotes.

(c) An implicit equation for the hyperbola.

10.

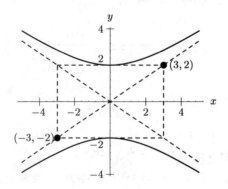

Figure 12.7

ANSWER:

(a) (0,2), (0,-2), (0,0)

(b) $y = \frac{2}{3}x, \ y = -\frac{2}{3}x$

(c) $\dfrac{y^2}{4} - \dfrac{x^2}{9} = 1$

11.

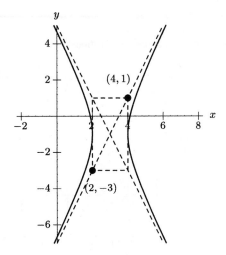

Figure 12.8

ANSWER:

(a) (2,-1), (4,-1), (3,-1)

(b) We need a line through (2,-3) and (4,1):

$$m = \frac{1 - (-3)}{4 - 2} = 2$$

Thus

$$y - 1 = 2(x - 4)$$
$$y = 2x - 7$$

Similarly, for the second line, use the points (2,1) and (4,-3):

$$m = \frac{1 - (-3)}{2 - 4} = -2$$

Thus

$$y - 1 = -2(x - 2)$$
$$y = -2x + 5$$

(c) $\dfrac{(x - 3)^2}{1} - \dfrac{(y + 1)^2}{4} = 1$

12. Write the parametric equation for the top branch of the hyperbola $y^2 - x^2 = 1$.
   ANSWER:
   $x = \sinh t, y = \cosh t$.

13. Write a parameterization for the right half of the hyperbola

$$\frac{(x - 2)^2}{4} - \frac{(y - 1)^2}{9} = 1$$

using hyperbolic functions.
   ANSWER:
   $x = 2 + 2\cosh t, \quad y = 1 + 3\sinh t, \quad -\infty < t < \infty$

14. Write a parameterization for the lower half of the hyperbola

$$2(y + 2)^2 - 3(x - 3)^2 = 12$$

using hyperbolic functions.
   ANSWER:

$$2(y + 2)^2 - 3(x - 3)^2 = 12$$
$$\frac{(y + 2)^2}{6} - \frac{(x - 3)^2}{4} = 1$$

$$x = -2 + 2\sinh t, \quad y = 3 - \sqrt{6}\cosh t, \quad -\infty < t < \infty$$

Problems

15. Describe the graph given by the following parametric equations.

$$x = \cos\left(\frac{\pi}{t+1}\right) \quad y = \sin\left(\frac{\pi}{t+1}\right)$$

ANSWER:
Table 12.1 shows values of x and y for various values of t.

Table 12.1

t	0	1	2	3	5	100
x	-1	0	1/2	$\sqrt{2}/2$	$\sqrt{3}/2$	≈ 1
y	0	1	$\sqrt{3}/2$	$\sqrt{2}/2$	1/2	≈ 0

The parameterization traces out the upper half of the unit circle in the clockwise direction. It never quite reaches the point (1,0).

16. Let $y = t$ be one part of a parametric equation. Find x as a function of t on $0 \le t \le 2$ which would complete the parameterization of the graph of a straight line from $(0,0)$ to $(1,1)$ and then continuing from $(1,1)$ in a straight line to $(0,2)$.

ANSWER:
$$x = \begin{cases} t & 0 \le t < 1, \\ 2 - t & 1 \le t \le 2. \end{cases}$$

17. A mouse hanging on to the end of the windmill blade in Figure 12.9 has coordinates given by

$$x = A\cos(kt) \quad y = A\sin(kt)$$

where the origin is at the center of the blades, x and y are in meters and t is in seconds. The blades are 5 meters long and the windmill makes one complete revolution every 24 seconds in a counterclockwise direction. The mouse starts in the 3 o'clock position and 9 seconds later, loses its hold and flies off.

(a) Find A and k.
(b) Find the speed of the mouse when it is on the blade.
(c) What is the angle between the blade and the positive x-axis at the moment the mouse flies off?
(d) Assume that when the mouse leaves the blade it moves along a straight line tangent to the circle on which it was previously moving. What is the equation of that line?

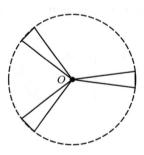

Figure 12.9

ANSWER:

(a) A is the amplitude, so $A = 5$ meters.
   k is given by the period $= \frac{2\pi}{k} = 24$ seconds, so $k = \frac{\pi}{12}$.
(b) The mouse travels through $2\pi \cdot 5 = 10\pi$ meters in one revolution, which takes 24 seconds. Thus, the mouse's speed is $\frac{10\pi}{24} = \frac{5\pi}{12}$ meters/sec.
(c) The blade is half-way between the 12 o'clock and the 9 o'clock positions, so the angle is $\frac{3\pi}{4}$ or $135°$ with the positive x-axis.

(d) The mouse leaves the blade at the point

$$x = 5\cos\left(\frac{9\pi}{12}\right) = \frac{-5}{\sqrt{2}} \qquad y = 5\sin\left(\frac{9\pi}{12}\right) = \frac{5}{\sqrt{2}}.$$

The line along which the mouse moves is perpendicular to the blade and, therefore, makes an angle of $45°$ with the positive x-axis. Thus, its slope is 1 and its equation is

$$y - \frac{5}{\sqrt{2}} = 1\left(x + \frac{5}{\sqrt{2}}\right)$$

$$y = x + 5\sqrt{2}$$

18. Put the following circle in standard from, and then find parametric equations for it.

$$x^2 + y^2 + 6x - 2y + 6 = 0$$

ANSWER:

$$x^2 + y^2 + 6x - 2y + 6 = 0$$
$$x^2 + 6x + 9 + y^2 - 2y + 1 = 4$$
$$(x + 3)^2 + (y - 1)^2 = 4$$
$$\frac{(x + 3)^2}{4} + \frac{(y - 1)^2}{4} = 1$$

For the parametric equations, $x = -3 + 2\cos t$, $y = 1 + 2\sin t$, $0 \le t \le 2\pi$.

19. (a) Eliminate the parameter and write an explicit formula for the curve $x = t^3 + 1, y = t^2 + 2$
   (b) Describe the curve as a transformation of a power function.
   ANSWER:

   (a) $y = (x - 1)^{2/3} + 2$.
   (b) The graph looks like $y = x^{2/3}$ translated 1 unit to the right and 2 units up.

20. (a) Eliminate the parameter and write an explicit formula for the curve $x = e^{0.5t}, y = e^t, 0 \le t \le 1$.
   (b) Find the endpoints of the curve.
   ANSWER:

   (a) $y = x^2$
   (b) $(1, 1)$ and $(\sqrt{e}, e)$

21. Find all points of intersection of the graphs of the following equations:

$$x^2 + y^2 = \frac{1}{2}$$
$$x^2 - y^2 = \frac{1}{4}$$

ANSWER:
Adding the equations gives

$$2x^2 = \frac{3}{4}$$

$$x = \pm\sqrt{\frac{3}{8}}$$

Substituting into the first equation gives

$$\frac{3}{8} + y^2 = \frac{1}{2}$$

$$y = \pm\sqrt{\frac{1}{8}}.$$

Since any pair of x and y values go together, the points are

$$\left(\sqrt{\frac{3}{8}}, \sqrt{\frac{1}{8}}\right); \left(\sqrt{\frac{3}{8}}, -\sqrt{\frac{1}{8}}\right); \left(-\sqrt{\frac{3}{8}}, \sqrt{\frac{1}{8}}\right); \left(-\sqrt{\frac{3}{8}}, -\sqrt{\frac{1}{8}}\right).$$

22. By completing the square, rewrite the equation for an ellipse $x^2 + 4y^2 - 4x - 8y - 8 = 0$ in the form

$$\frac{(x-h)^2}{a^2} + \frac{(y-k)^2}{b^2} = 1.$$

What is the center of the ellipse?
ANSWER:

$$x^2 + 4y^2 - 4x - 8y - 8 = 0$$
$$x^2 - 4x + 4 + 4(y^2 - 2y + 1) = 16$$
$$(x-2)^2 + 4(y-1)^2 = 16$$
$$\frac{(x-2)^2}{16} + \frac{(y-1)^2}{4} = 1$$

The center is at $(2,1)$.

23. Write parametric equations for the quarter of an ellipse centered at $(0,0)$, starting at $(0,6)$ and ending at $(-3,0)$.
ANSWER:
$x = 3\cos t, y = 6\sin t, \frac{\pi}{2} \le t \le \pi$.

24. Find all intersections of the ellipse $\frac{(x+3)^2}{4} + \frac{(y-4)^2}{2} = 1$ with the line $x = -2$.
ANSWER:
In the equation of the ellipse, let $x = -2$ and solve for y. The two solutions give

$$\left(-2, \frac{8+\sqrt{6}}{2}\right) \text{ and } \left(-2, \frac{8-\sqrt{6}}{2}\right)$$

25. P is a point on the unit circle, Q and R lie on the ellipse in Fig 12.10. If the equation of the ellipse is given by the function

$$\frac{x^2}{4} + y^2 = 1$$

(a) Find the y-coordinate of point R in terms of θ.
(b) Find the x-coordiante of point Q in terms of θ.

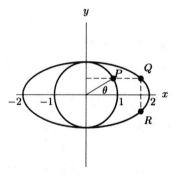

Figure 12.10

ANSWER:
(a) The y-coordinate of R is $(-1)\cdot(y$-coordinate of $Q)$. But the y-coordiante of Q is equal to the y-coordinate of P and the y-coordiante of P is $\sin\theta$. Thus the y-coordinate of R is $-\sin\theta$.
(b) We know that the equation of the ellipse is given by

$$\frac{x^2}{4} + y^2 = 1$$

Since we know that the y-coordinate of Q is $\sin\theta$ we know that the x-coordinate must satisfy the equation of the ellipse, or in other words

$$\frac{x^2}{4} + \sin^2\theta = 1$$

Solving we get

$$\frac{x^2}{4} + \sin^2\theta = 1$$

$$\frac{x^2}{4} = 1 - \sin^2\theta$$

$$= \cos^2\theta$$

$$x^2 = 4\cos^2\theta = (2\cos\theta)^2$$

$$x = \pm 2\cos\theta$$

We know that the x-coordinate of Q is greater than 0 and $0 < \theta < 90°$ so the x-coordinate of Q is $2\cos\theta$.

26. Parameterize the hyperbola in Exercise 10. What t values give the upper half?

ANSWER:

$x = 3\tan t$, $y = 2\sec t$. To see which values of t give the upper half, construct a table of values

Table 12.2

t	0	$\frac{\pi}{4}$	$\frac{3\pi}{4}$	π	$\frac{5\pi}{4}$	$\frac{7\pi}{4}$
x	0	3	-3	0	-3	3
y	2	$2\sqrt{2}$	$-2\sqrt{2}$	-2	$-2\sqrt{2}$	$2\sqrt{2}$

The upper half occurs when $0 \le t \le \frac{\pi}{2}$, $\frac{3}{2}\pi \le t \le 2\pi$.

27. Parameterize the hyperbola in Exercise 11. What t values give the left half?

ANSWER:

$x = 3 + \sec t$, $y = -1 + 2\tan t$. To see which values of t give the left half, construct a table of values

Table 12.3

t	0	$\frac{\pi}{4}$	$\frac{3\pi}{4}$	π	$\frac{5\pi}{4}$	$\frac{7\pi}{4}$
x	4	$3 + \sqrt{2}$	$3 - \sqrt{2}$	2	$3 - \sqrt{2}$	$3 + \sqrt{2}$
y	-1	1	-3	-1	1	-3

The left half occurs when $\frac{\pi}{2} \le t \le \frac{3}{2}\pi$.

28. By completing the square, rewrite the equation $-9x^2 + 4y^2 - 36x - 16y - 56 = 0$ in the form

$$\frac{(x-h)^2}{a^2} - \frac{(y-k)^2}{b^2} = 1 \quad \text{or} \quad \frac{(y-k)^2}{b^2} - \frac{(x-h)^2}{a^2} = 1.$$

What is the center, and does the hyperbola open left-right or up-down?

ANSWER:

$$-9x^2 + 4y^2 - 36x - 16y - 56 = 0$$

$$4y^2 - 16y + 16 - 9x^2 - 36x - 36 = 36$$

$$4(y-2)^2 - 9(x+2)^2 = 36$$

$$\frac{(y-2)^2}{9} - \frac{(x+2)^2}{4} = 1.$$

The center is $(-2, 2)$, and the parabola opens up-down.

29. On the same set of axes, graph the hyperbola

$$\frac{x^2}{n^2} - \frac{y^2}{n^2} = 1$$

for $n = 1, 2, 3$. Describe in words what happens to the graphs as n increases.

ANSWER:

The vertices get farther apart as n increases, but the asymptotes remain the same.

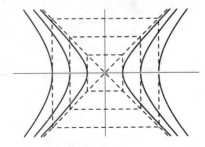

Figure 12.11

30. Show that $f(x) = \cosh x$ is an even function and that $g(x) = \sinh x$ is an odd function.
  ANSWER:
  $f(-x) = \frac{e^{-x}+e^x}{2} = f(x)$ and $g(-x) = \frac{e^{-x}-e^x}{2} = -g(x)$.

31. Factor $(\cosh^2 x - \sinh^2 x)$ to give an alternate proof to the identity
$$\cosh^2 x - \sinh^2 x = 1.$$

ANSWER:

$$(\cosh^2 x - \sinh^2 x) = (\cosh x + \sinh x)(\cosh x - \sinh x)$$
$$= \left(\frac{e^x + e^{-x}}{2} + \frac{e^x - e^{-x}}{2}\right)\left(\frac{e^x + e^{-x}}{2} - \frac{e^x - e^{-x}}{2}\right)$$
$$= (e^x)(e^{-x})$$
$$= e^0$$
$$= 1$$

32. The function $\sinh x$ is called the *hyperbolic sine*, and is given by the formula
$$\sinh x = \frac{1}{2}(e^x - e^{-x})$$

One of the following four functions is the formula for the inverse function $\sinh^{-1} x$ (on a suitable domain). Decide which formula is correct, and show the calculations which justify your answer. (Note that you are not being asked to solve for the inverse function, only to check which of the following four possibilities is correct.)

(a) $y = \frac{1}{2}\ln(\frac{1+x}{1-x})$ (b) $y = \frac{1}{2}\ln(\frac{x+1}{x-1})$

(c) $y = \ln(x + \sqrt{x^2 + 1})$ (d) $y = \ln(x + \sqrt{x^2 - 1})$

ANSWER:
  The inverse function of $f(x)$ must satisfy $f(f^{-1}(x)) = x$. Therefore, by substituting each of the four formulas into $f(x) = \sinh x$, we can see which is the inverse. For formula (c), we see that it does give x:

$$x = \frac{1}{2}(e^y - e^{-y})$$
$$= \frac{1}{2}\left(e^{\ln(x+\sqrt{x^2+1})} - e^{-\ln(x+\sqrt{x^2+1})}\right) \quad \text{Substituting } y = \ln(x + \sqrt{x^2 + 1})$$
$$= \frac{1}{2}\left(x + \sqrt{x^2 + 1} - \frac{1}{x + \sqrt{x^2 + 1}}\right) \quad \text{Combining fractions}$$
$$= \frac{1}{2}\left(\frac{(x + \sqrt{x^2 + 1})^2 - 1}{x + \sqrt{x^2 + 1}}\right)$$
$$= \frac{1}{2}\left(\frac{x^2 + 2x\sqrt{x^2 + 1} + x^2 + 1 - 1}{x + \sqrt{x^2 + 1}}\right)$$
$$= \frac{1}{2}\left(\frac{2x^+2x\sqrt{x^2 + 1}}{x + \sqrt{x^2 + 1}}\right)$$
$$= \frac{1}{2}\left(\frac{2x(x + \sqrt{x^2 + 1})}{x + \sqrt{x^2 + 1}}\right)$$
$$= x$$